Fundamentals of Sustainable Urban Design

Avi Friedman

Fundamentals of Sustainable Urban Design

Springer

Avi Friedman
McGill University School of Architecture
Montreal, QC, Canada

ISBN 978-3-030-60864-4 ISBN 978-3-030-60865-1 (eBook)
https://doi.org/10.1007/978-3-030-60865-1

This Springer imprint is published by the registered company Springer Nature Switzerland AG
The registered company address is: Gewerbestrasse 11, 6330 Cham, Switzerland

Preface

Fundamental global and local social changes bring about a need to rethink urban design strategies and align them along sustainable principles. Past approaches no longer sustain new demands of innovation. The need for a new outlook is propelled by environmental, social, and economic transformations in large and small communities.

The depletion of nonrenewable natural resources, elevated levels of greenhouse gas emissions, and climate change are a few of the environmental challenges. They force urban designers to reconsider conceptual approaches in favor of ones that promote a better suitability between communities and nature. Consideration of overall planning concepts that minimize a development's carbon footprint by including district heating, considering passive solar gain, promoting local renewable energy production, and preserving the site's natural assets are some strategies that architects and builders are integrating into their thought process and practice.

Social challenges are also drawing the attention of key stockholders. As the "baby-boom" generation is retiring, accommodating elderly populations in cities and homes is taking priority. Walkable communities, aging in place, and multigenerational living are some of the concepts considered. In addition, live-work environments have become part of the economic reality for those who wish to work from home—made possible through digital advances. In addition, a call to reduce urban sprawl, shrinking family size, and rapidly growing aging populations have led to demands for new housing types.

Economic fluctuations and shifting buying and production trends have affected world markets and the lives of individuals leading to a downturn in some cities' wealth and the creation of new types of business hubs in others. For one, the rise of grassroots and sharable initiatives seems to be a natural outcome of these transformations. Establishing an economy somewhat independent of global and national trends has always been an aim of many communities, but recent crises may increase interest in this goal considerably. Cities are resorting to measures that encourage people to use local products and services.

As a result of these transformations, this book is aiming to offer sustainable urban design strategies and practices for the planning of new cities and renewal of existing ones, to make them resilient to the present and future challenges. The need to think innovatively and systematically about the design of cities also guided the structure and the content of this book that is organized in four parts.

Part I, Adapting to Changing Times, sets the stage for all the book's chapters. The subjects discussed in this part are Chap. 1: Present and Future Challenges, Chap. 2: Sustainability and Urban Design.

Part II, Urban Forms and Buildings' Shape, examines the broader planning issues of cities. It includes Chap. 3: Master Planning for Sustainability, Chap. 4: The Form of a Place, Chap. 5: Choosing a Suitable Density, Chap. 6: Orienting for the Elements, Chap. 7: Space Defining Buildings, Chap. 8: Sense of Place, Human Scale, and Vistas, Chap. 9: Mixing Land Uses and Dwelling Types, Chap. 10: Infill Projects and Strategies for their Integration, Chap. 11: Urban Design for Growth and Change, Chap. 12: Identity and Diversity of Districts and Buildings, Chap. 13: Planning for Energy Distribution and Waste Collection, Chap. 14: Communities with a Digital Heart.

Part III deals with Mobility and Connectivity in Cities. It includes Chap. 15: Mobility and the City; the Broad View, Chap. 16: Urban Design for Transit-Oriented Development, Chap. 17: Alternative Standards for Streets, Paths, and Pavements, Chap. 18: Urban Design for Safe Walking and Biking, Chap. 19: Car-Free Environments and Shared Streets, Chap. 20: Public Transit and Urban Design, Chap. 21: Urban Design and Shared Transport, Chap. 22: Accommodating Seniors and People with Reduced Mobility, Chap. 23: Accessibility and Livability in Winter Cities.

Part IV deals with Public and Green Open Spaces. It includes Chap. 24: Open Spaces as an Urban System, Chap. 25: Integrating Existing Natural Features, Chap. 26: Urban Design for Biodiversity, Chap. 27: Planting and Landscaping for Sustainability, Chap. 28: Open Spaces for Healthy Living, Chap. 29: Urban Agriculture and Community Gardens, Chap. 30: Urban Design for Social Engagement, and Chap. 31: Public Art and Street Furniture.

The trust of the book lays in the fact that sustainability is the overarching aspect of all the strategies and practices discussed. It offers a systematic and structured approach to urban design to cover aspects central to decision-making. The theories are accompanied by cases and illustrations that support the concepts described. When put together, these subjects offer a comprehensive compendium of issues that will guide urban design in changing times.

Montreal, QC, Canada — Avi Friedman

Acknowledgments

Sustainable planning was a subject of my work for many years. It included collaboration with and contribution by numerous colleagues, assistants, and students who directly and indirectly inspired my work. My apology if I have mistakenly omitted the name of someone who contributed to this book. I will do my best to correct such omission in future editions.

This book could not have been written without contribution to the background research, compiling information, and the writing by a team of highly dedicated assistance. It included my outstanding former students Emmanuelle (Emma) Bandia, Simone Dayal-Escutin, Alexandra Pollock, and Kelly Waldron. Their dedication, hard work, talent, and punctuality are most appreciated.

Assembling the book's material, its organization, and editing was coordinated by another former student Genessa Bates. I truly appreciate Genessa's hard work, dedication, insistence on clarity, accuracy, and outstanding communication skills.

Special thanks are extended to Charles Grégoire, Elisa Costa, Jeff Jerome, Zhong Cai, Diana Nigmatullina, Josie White, Juan Mesa, Jing Han (Jay), David Auerdach, Isabella Rubial, Rainier Silva, JJ Zhao, Xiong Wu Fa, Amelie Lessard, Nyd Garavito-Bruhn, Na Zhang, and Jing Yan Liu who drew the illustrations. Their talent and insistence on achieving excellence is truly appreciated and admired.

Many thanks are also extended to the elected officials and administrators of the cities who invited me to consult them and are featured in the book. My appreciations go to my design team members, who are listed in the Projects' Teams list for their utmost dedication.

To Michael Luby, Senior Publishing Editor at Springer and Nicole Lowary, Assistant Editor, many thanks for the trust and the guidance.

Finally, my heartfelt thanks and appreciation to my wife Sorel Friedman, Ph.D., and children Paloma and Ben for their love and support.

Contents

Part I Adapting to Changing Times

1 **Present and Future Challenges** 3

2 **Sustainability and Urban Design** 15

Part II Urban Forms and Buildings' Shape

3 **Master Planning for Sustainability** 29

4 **The Form of a Place** 37

5 **Choosing a Suitable Density** 49

6 **Orienting for the Elements** 57

7 **Space Defining Buildings** 67

8 **Sense of Place, Human Scale, and Vistas** 75

9 **Mixing Land Uses and Dwelling Types** 85

10 **Infill Projects and Strategies for Their Integration** 95

11 **Urban Design for Growth and Change** 105

12 **Identity and Diversity of Districts and Buildings** 115

13 **Planning for Energy Distribution and Waste Collection** 125

14 **Communities with a Digital Heart** 135

Part III Mobility and Connectivity

15 **Mobility and the City: The Broad View** 145

16 **Urban Design for Transit-Oriented Development** 155

17 Alternative Standards for Streets, Paths, and Pavements 163

18 Urban Design for Safe Walking and Biking 171

19 Car-Free Environments and Shared Streets 181

20 Public Transit and Urban Design 187

21 Urban Design and Shared Transport 197

22 Accommodating Seniors and People with Reduced Mobility 205

23 Accessibility and Livability in Winter Cities 213

Part IV Public and Green Open Spaces

24 Open Spaces as an Urban System 225

25 Integrating Existing Natural Features 235

26 Urban Design for Biodiversity 245

27 Planting and Landscaping for Sustainability 257

28 Open Spaces for Healthy Living 267

29 Urban Agriculture and Community Gardens 277

30 Urban Design for Social Engagement 285

31 Public Art and Street Furniture 295

Illustration Credits 307

Projects' Teams 309

Bibliography 311

Index 321

Part I
Adapting to Changing Times

Chapter 1
Present and Future Challenges

Abstract Contemporary social, economic, and environmental challenges that affect the built environment require an innovative approach when planning cities. This means that current practices must change, yet, it is vital that these changes be understood as an evolution—a transition with a direct connection to the past. This chapter will clarify this connection. First, a brief history of what lead to the decline of cities will be given, followed by an exploration of social issues that are imperative to consider.

Keywords Climate change · Traffic congestion · Public health · Aging population · Demographic changes · Social infrastructure · Urban sprawl · Carbon dioxide emissions · Affordability gap · New Urban Agenda · Green growth · Waste disposal

1.1 The Decline of Sustainable Planning

Historically, the way cities were built were, in part, inherently sustainable—without the idea of sustainability being a recognized concept at the time. Old settlements were pedestrian, and citizens primarily traveled by foot in narrow streets (Fig. 1.1). Some cities have even maintained this role, in cases where the topography does not allow for anything else, or where economic and social activities continue to support foot traffic (Gehl 2010). For example, medieval cities—due to their compact nature—were easily walkable and supported trade and craftsmanship due to this high mobility (Gehl 2010).

The progressive and severe decline of the aforementioned sustainable planning can be identified in the decades following WWII. This period saw a drastic change in the way cities were planned and lived in. Planning prioritized the development of grand, individual buildings which served separate purposes, rather than mixed-use areas and conglomerations of city spaces as a whole (Gehl 2010). The explosive use of vehicles is a phenomenon which coincides with both the decline in sustainable planning and the expansion of the modernist movement. Roads cut through city spaces, separating public and pedestrian areas. Increased traffic lead to lifeless communities devoid of people, where vehicular mobility was prioritized overactive mobility and open public space. This also coincided with the growth of the suburbs which is characterized by mass-produced housing units, increased use of automobiles, and a highly privatized

A. Friedman, *Fundamentals of Sustainable Urban Design*,
https://doi.org/10.1007/978-3-030-60865-1_1

Fig. 1.1 Due to their small footprint, old cities with narrow roads were highly walkable

landscapes which led to urban sprawl in many nations (Fig. 1.2). There are several notable examples of this model of suburban development, but perhaps the most remarkable example is Levittown, a development constructed in Long Island, United States of America (USA) between 1947 and 1951 (Ruff 2007).

Fig. 1.2 Since the middle of the twentieth century many cities such as Shanghai, China (left) and Queretaro, Mexico (right) experienced rapid urban sprawl

Fig. 1.3 Proper human scale and a mix of residential and commercial land uses enhanced walkability in old cities such as Saint-Valery-sur-Somme, France

Jane Jacobs—a journalist and prominent writer and thinker on urban studies—was a catalyst in the movement against modernist planning principles following the release of her book *Death and Life of Great American Cities* (1961). She claimed that the modernist movement and the infrastructures inspired by it rejected city space, gradually breaking down public meeting places, instead creating spaces which streamlined activity to specific purposes (Gehl 2010). Throughout her career, Jacobs advocated for a shift in planning toward more sustainable principles, prioritizing people and mixed-use public spaces (Fig. 1.3).

Prominent architect and urban scholar Jan Gehl studies human scale, public spaces and how people use the public domain. This is a fundamental concept which was considered when cities were built for pedestrians, not vehicles; when buildings did not reach heights exceeding a comfortable gaze and roads were not wide enough to fit multiple cars; in these cities people and human interactions were central to the local economy and social life (Gehl 2010).

It is perhaps today, after the wave of modernism and unsustainable practices, that the importance of sustainability has been recognized and become particularly relevant in urban planning; there is a growing consensus that current practices must change. Development of single-family dwellings only connected by roads, such as suburbs, has become too costly and isolates residents from each other. A new type of development should be prioritized for future planning. To do so, past cases of successful urban design must be studied in order to identify methods which are proven to have a positive impact in cities. It is equally important to look forward and foresee the new challenges that planners will face.

1.2 Contemporary Urban Challenges

Globally, the proportion of people living in cities continues to rise, putting enormous pressure on already dense urban spaces. This pressure is exacerbating existing social challenges such as climate change, traffic congestion, public health, and an aging population. Social, environmental and economic changes are transforming the places we live in, and policies and strategies must be introduced in order to adapt and build in resiliency.

It should be noted that variations exist in the prevalence of these issues between continents, nations, and cities. The specific challenges experienced in a city are unique to that place and the solutions required should be developed by considering the local context. The work that follows aims to present solutions to issues commonly experienced in urban spaces today.

1.2.1 Social Transformations

Firstly, significant demographic changes are occurring in urban populations. The proportion of citizens age 55 or above is rapidly increasing in Western countries. The generation referred to as the baby boomers, born between 1946 and 1965, was 35.5% of the Canadian population in 2018 and is rapidly increasing as illustrated in Fig. 1.4 (Statistics Canada 2019). The increase in an aging population coupled with a shrinking youth population is expected to strain the provision of public services; it will be necessary to adapt infrastructure and public services in order to address the needs of this cohort. These adaptations can also offer an opportunity for communities to implement sustainable health and social infrastructure for future generations.

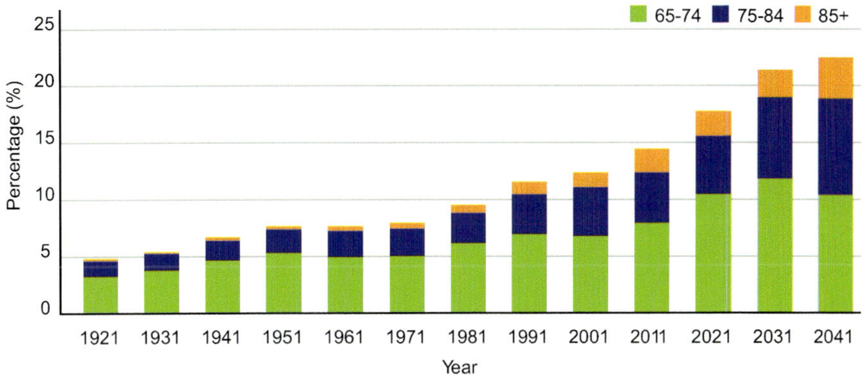

Fig. 1.4 The aging population of Canada and many other nations will increase in the coming decades requiring introduction of urban design measures to accommodate seniors' needs

Another important demographic trend is the decline in the size of the average household. For example, in Canada the average household size in 1941 was 4.3 persons, in 2011 the average was 2.5 persons, and the average is continuing to decrease (Statistics Canada 2018). Coupled with the increasing cost of living in many urban areas, this has led to an increased demand for homes and apartments of smaller size.

1.2.2 Environment Challenges

Within the environmental domain there are a few changes which will have consequence for how people live in cities. There is a clear connection between economic development, urban form and their environmental implications. Thus, in order to reduce negative environmental impacts, incorporating and considering environmental factors in urban policy is crucial. Urban sprawl is linked to a number of negative environmental impacts including, but not limited to: "loss of environmentally fragile lands, reduced regional open space, greater air pollution, higher energy consumption, loss of farmland, reduced diversity of species, increased runoff of stormwater, increased risk of flooding, excessive removal of native vegetation, and ecosystem fragmentation" (Johnson 2001).

The negative environmental impacts of urban sprawl and single-family home development which increases the carbon dioxide emissions have given rise to a need for densification. The extensive use of private vehicles is characteristic of low-density neighborhoods and demonstrative of unsustainable, energy-intensive means of transportation (Fig. 1.5). Another characteristic is the use of zoning policies to separate spaces by function, meaning citizens must travel longer distances between activities, increasing dependence on vehicles (Johnson 2001). When a development is dense and mixed-use, residents share resources and do not have to travel long distances to use different services. This form of development largely resembles premodernist

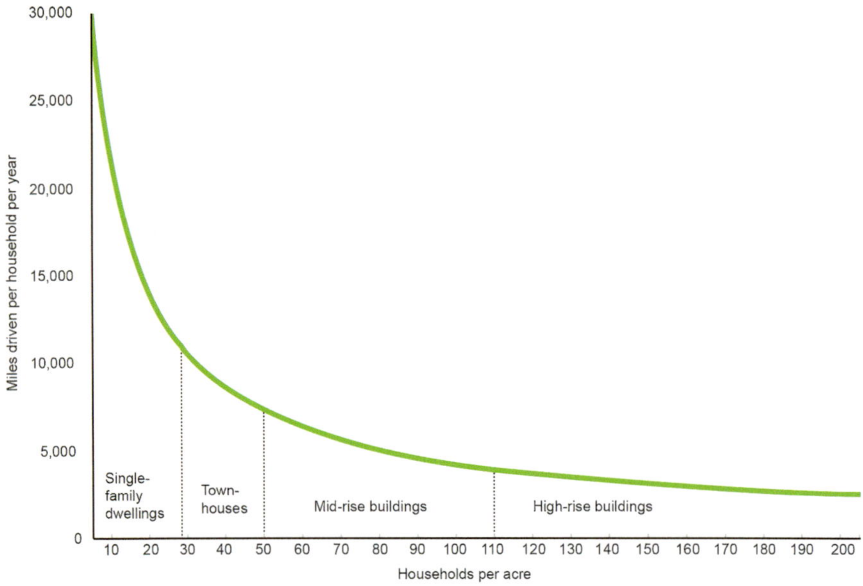

Fig. 1.5 Residents living in US communities with higher densities tend to drive three times less than those who live in neighborhoods with single-family homes

urban planning, designing urban spaces to accommodate pedestrians rather than cars. Within this model, residents live in close proximity to a variety of shops and services and walk or use public transit to travel.

Other environmental issues pose a threat to urban spaces; rising sea levels, floods, extreme weather events, and earthquakes can all threaten the safety of residents. Due to the increase of floods and other unpredictable climatic phenomena, adapting to environmental changes is a necessity. Cities should be planned with systems in place to protect residents from these environmental factors.

1.2.3 New Economic Realities

Changes in the contemporary economic landscape must be taken into consideration. Commerce has evolved, economic activities around the world are increasingly interconnected, blurring the line between global and local. Companies no longer need to be tied to a location and can organize operations from afar using digital communication (Friedman 2017). This has had a ripple effect on the nature of commerce, namely with the rise of online shopping and the closing of traditional main street commerce (Fig. 1.6). It is difficult to predict what kind of shopping experience consumers will prefer in the future, but consumption trends will shape communities and their needs in relation to their immediate built environment.

Relating to the affordability of housing in cities, new economic and social realities also put the single-family detached home beyond the reach of a growing number of

Fig. 1.6 Many cities like Lancaster, UK saw store closing a result of online shopping

first-time home buyers in many nations. There is an increased demand for less expensive, smaller units as a result of the increased cost of living and shrinking family sizes. An affordability gap has emerged, where the rate of increase of housing prices has far surpassed the increase of family incomes. This widening gulf in affordability in some regions can be explained by higher land and infrastructure costs, offering another argument for densification and the building of affordable neighborhoods.

To conclude, cities today are rapidly changing as populations evolve and as technological advancements are made. These cities are facing a new set of social, environmental, and economic challenges. It is necessary to adapt and look at possible urban design solutions, through new developments while also learning from best past practices.

1.3 The Emergence of the Green City Thinking

As the aforementioned challenges in cities become increasingly relevant and continue to persist, the emergence of green city thinking is on the rise. This is apparent in cities around the world, under initiatives and movements of many different forms.

Fig. 1.7 Many of Singapore's buildings' have vegetation growing on their facades

For instance in Singapore, green building became mandatory in 2008, setting higher environmental standards for new buildings as illustrated in Fig. 1.7 (Building and Construction Authority 2019); in Luxembourg it was announced that in 2019 all public transport would become free of charge (Ville de Luxembourg 2019); and in Bogotá, Colombia, on each Sunday and public holidays certain main streets are blocked off to cars, creating a 120 km (74.6 miles) network of safe lanes for cyclists and pedestrians under a program called Ciclovía (Instituto Distrital de Recreación y Deporte 2019).

Green thinking has not only become a popular trend and an economic opportunity, but more importantly formed a basis for political discourse, formal planning, and investments in cities. At the 2016 UN Habitat conference in Quito, a plan for building cities and promoting sustainable urban development was created with input from all UN Member States and relevant stakeholders called the New Urban Agenda. The framework established a model and set of guidelines to promote sustainable urban development and recognizes the importance of targeting contemporary challenges facing cities today. Some of the main challenges identified were climate change, promoting the conservation and sustainable use of water, promoting environmentally sound waste management, and sustainable use of natural resources (United Nations 2017).

Amidst the positive discourse and environmental commitments and movements, it is important to define what a green city is and identify its main priorities. The Asian Development Bank (Lindfield and Steinberg 2012) defines green cities as cities that have already achieved, or working toward long-term environmental sustainability, and distinguishes these from cities which continue to follow environmentally unsustainable trajectories. The Organization for Economic Co-Operation and Development, in its green cities program, defines green growth as a "new paradigm that promotes economic development while reducing greenhouse gas emissions and pollution, minimizing waste and inefficient use of natural resources and maintaining biodiversity" (OECD 2013). Finally, Cohen offers a definition of the "sustainable city," highlighting a number of key environmental aspects. He states that the sustainable city uses as few nonrenewable resources as possible, uses energy and water as efficiently as possible, and attempts to reduce and recycle

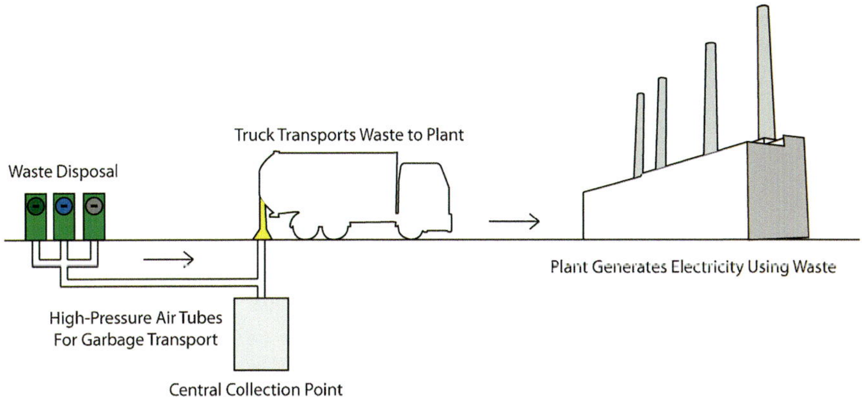

Fig. 1.8 A diagram showing the waste collection system for energy production in Hammarby near Stockholm, Sweden

waste, minimizing the impact of waste disposal, as is the case in Hammarby, Sweden (Fig. 1.8) (Cohen 2018). Although the focus of each definition varies, in each case putting environmental sustainability first is a priority. In a time of innovation, change, and increasing circulation of many green ideas, it is important to identify which investments are vital for creating green cities and optimizing environmental sustainability.

1.3.1 Key Investments in Green City Thinking

There are several key investments which ensure the successful development of a green city. Firstly, Transit-Oriented Development (TOD), defined by Belzer and Autler (2002) as "intense, comprehensive development around transit stations" is an important investment. TOD should be "mixed-use, walkable, location-efficient development that balances the need for sufficient density to support convenient transit service with the scale of the adjacent community" (Belzer and Autler 2002). This form of development is instrumental in reducing the negative impacts of single-driver vehicles, as well as creating higher-density communities that are located near transit corridors where walkability, bicycle culture, and mixed land use form key planning principles. Investing in the necessary infrastructure for such development is expensive, but beneficial for the long-term sustainability of a community as was the case in Copenhagen's urban plan (Fig. 1.9). We will explore the history and implementation of TOD further in Sect. 16.1.

Achieving livable density is another key investment. This implies attaining a level of density within a space that combines inviting city spaces and a certain critical mass of people who want to use it (Gehl 2010). Moreover, a livable, dense neighborhood combines different types of land use in the same area, without

Fig. 1.9 The regional plan of Copenhagen, Denmark designated five corridors of urban development with highly efficient transit services

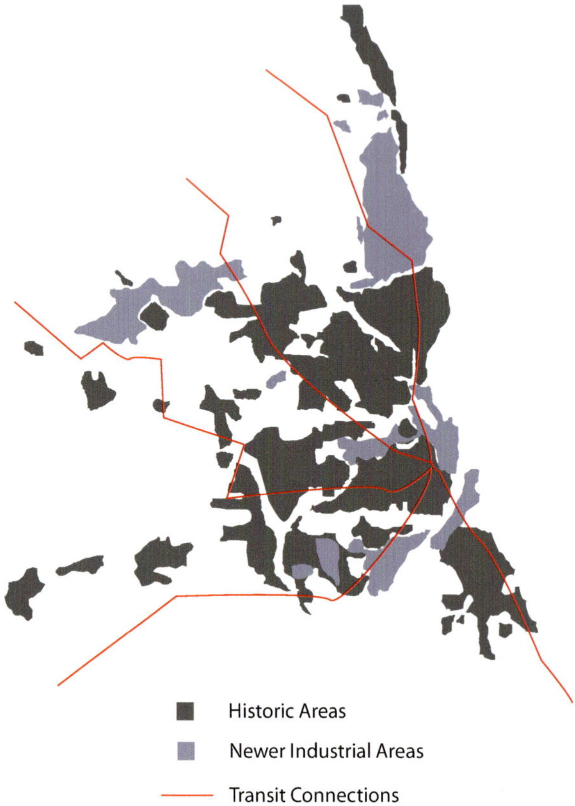

comprising on livability, and maintaining a human scale. Creating dense neighborhoods goes hand-in-hand with TOD: as resources necessary for economic activity tend to be channeled into areas in which physical access is greatest, the planning and design of transportation networks greatly impacts the spatial dispersion of urban development (Lindfield and Steinberg 2012). In other words, the density of neighborhoods is greatly determined by the structure of transport networks within and between them. As elaborated in Sect. 1.1, dense, mixed-use developments have ample benefits, including reducing per capita consumption of energy and resources and promoting active mobility by reducing travel distances to various services.

Subsequently, investment in green technology is also important in the development of a green city. Particularly today, where there is great investment and innovation in scientific technologies which can improve environmental sustainability in many ways. This includes renewable technologies such as solar panels, improved insulation materials, smart heating regulation devices for the home, machines to sort waste and recycle bottles, along with other inventions (Fig. 1.10).

Lastly, city greening is the final investment emphasized in this chapter. It is the act of bringing more of nature into the city, thereby increasing the size and number

Fig. 1.10 Homes in a solar-powered community in Almere, the Netherlands

of green open spaces. Green areas improve a community's sustainability in many ways; they mitigate the urban heat island effect, increase biodiversity, improve air quality, and provide residents with a beautiful public space. In addition to increasing urban temperatures, the urban heat island effect can be harmful to communities by contributing to peak summer energy demands, air conditioning costs, air pollution, greenhouse gas emissions, heat-related illness, and water quality. Furthermore, having trees in a city provide shade in warm summer months, cool the air by absorbing rather than reflecting heat and decrease air pollution (Gehl 2010). Overall, trees, landscaping and flowers not only play an important environmental and health role; they also create inviting public spaces for citizens to spend more time outdoors, socialize and engage in healthy activities such as playing sports and exercising.

It is important to note that due to the widespread use of the term "green" in the context of urban development, the word often loses true sense of meaning; inappropriately being adopted in certain contexts, misrepresenting the environmental efforts of a company or other actor. In other words, this so-called "greenwashing" is a commonality, and recognizing the use of such is necessary when analyzing urban development projects. For example, the new master-planned development of Forest City in Malaysia advertises itself as a "low-carbon, green city, with a sustainable community incorporating a system of energy efficient buildings that meet green standards" (Forest City 2019). However, the entire development is dependent on massive amounts of concrete and is entirely built on reclaimed land which has disrupted the mangroves and native ecosystems. Although the proposed plans of this city include many plants and buildings covered in greenery, whether or not this development as a whole is truly environmentally sustainable should be reconsidered.

It must be recognized that cities are places of immense opportunity for change, social and economic progress and technological innovation. Overall, thanks to the high concentration of people and knowledge, cities provide a unique opportunity to adapt to changing times. If well designed, cities have the opportunity to be very green (Cervero et al. 2017). Where resources are used efficiently and when quality of life and human well-being are prioritized, cities become more pleasant and more sustainable places to live. In conclusion, the tools and investments needed to minimize negative environmental impact are known—it is a matter of carefully planning and keeping the green city in mind.

References

Belzer D, Autler G (2002) Transit oriented development: moving from rhetoric to reality. Brookings Institution Center on Urban and Metropolitan Policy, Washington, pp 6–15

Building and Construction Authority (2019) Regulatory requirements for new buildings and existing buildings undergoing major additions and alterations (A&A). https://www1.bca.gov.sg/regulatory-info/legislation-on-environmental-sustainability-for-buildings/regulatory-requirements-for-new-buildings-and-existing-buildings-undergoing-major-additions-and-alterations-(a-a). Accessed 4 Feb 2020

Cervero R, Guerra E, Al S (2017) Urban recalibration. In: Beyond mobility. Island Press, Washington, pp 1–14

Cohen S (2018) Defining the sustainable city. Columbia University Press, New York

Forest City (2019) Masterplan. https://forestcitycgpv.com/why-forest-city/masterplan/. Accessed 4 Feb 2020

Friedman A (2017) Designing sustainable communities. Bloomsbury, London

Gehl J (2010) Cities for people. Island Press, Washington

Instituto Distrital de Recreación y Deporte (2019) Ciclovía Bogotana. https://www.idrd.gov.co/ciclovia-bogotana. Accessed 4 Feb 2020

Jacobs J (1961) Death and life of great American cities. Vintage Books, New York

Johnson M (2001) Environmental impacts of urban sprawl: a survey of the literature and proposed research agenda. Environ Plann A 33(4):717–735

Lindfield M, Steinberg F (2012) Green cities. Asian Development Bank Institute. ProQuest Ebook Central, https://ebookcentral.proquest.com/lib/mcgill/detail.action?docID=3110769. Accessed 4 Feb 2020

OECD (2013) Green growth in cities. OECD green growth studies. OECD Publishing. https://doi.org/10.1787/9789264195325-en. Accessed 4 Feb 2020

Ruff J (2007) For sale: the American dream. World History Group, Virginia

Statistics Canada (2018) The shift to smaller households over the past century. https://www150.statcan.gc.ca/n1/pub/11-630-x/11-630-x2015008-eng.htm. Accessed 1 April 2020

Statistics Canada (2019) Analysis: population by age and sex. https://www150.statcan.gc.ca/n1/pub/91-215-x/2018002/sec2-eng.htm. Accessed 1 April 2020

United Nations (2017) The new urban agenda. http://habitat3.org/wp-content/uploads/NUA-English.pdf. Accessed 4 Feb 2020

Ville de Luxembourg (2019) Introduction of free public transport in Luxembourg. https://www.vdl.lu/en/getting-around/bus/tickets/free-travel. Accessed 04 Feb 2020

Chapter 2
Sustainability and Urban Design

Abstract A review of the social transformations that were outlined in Chap. 1 begs the question: how should the required changes be ushered in? how should communities be planned anew or retooled? Sustainability, as a philosophy and organizing principle, has been put forward as an answer to these questions. This chapter discusses sustainability in the context of urban design and offers key principles to evaluate and guide the process of making places sustainable.

Keywords Sustainable development · Social equity · Environmental sustainability · Governance · Sociopolitical · Social sustainability · Human rights · Cultural sustainability · Walkability · Cradle-to-cradle · Self-sustaining system · Symbiotic relationships · Life cycle approach

2.1 A Brief History of Sustainability

The origin of the term "sustainable development" can be traced back several decades. In 1972, the United Nations Conference on the Human Environment in Stockholm addressed concerns about the carrying capacity of the earth. The meeting served as a forum for the first international discussion on the relationship between ongoing environmental damage and the future of humanity. It was recognized then that population growth in some nations and overconsumption in others was causing ecological issues such as land degradation, deforestation, air pollution, and water scarcity.

Years later, this reflection led to the establishment of the international initiative World Commission on Environment and Development (WCED). In the 1987 report, *Our Common Future,* sustainable development was defined as a process that "meets the needs of the present without compromising the ability of future generations to meet their own needs" (Brundtland et al. 1987). This definition established a conceptual approach to sustainable development, whereby any action must be pursued with its future effects in mind. The WCED also created a paradigm for development whose main anchors are a need for social equity; incorporation of fair distribution of resources within and among nations; and the need to resolve conflicts caused by

A. Friedman, *Fundamentals of Sustainable Urban Design*,
https://doi.org/10.1007/978-3-030-60865-1_2

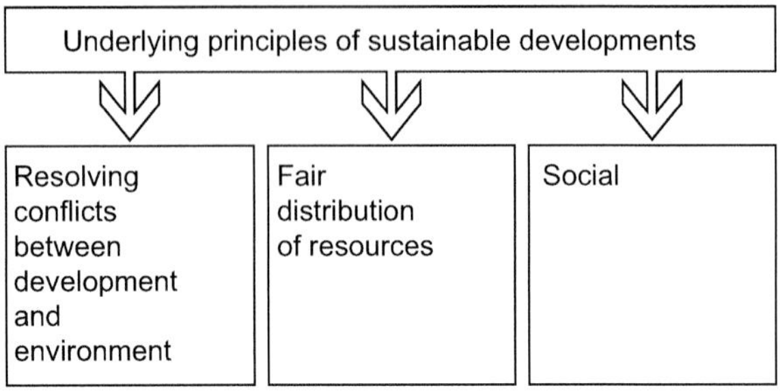

Fig. 2.1 The three underlying principles of sustainable developments according to authors of the 1987 *Our Common Future* report

development pressures and the environment (Fig. 2.1). These priorities have become a standard for sustainable development initiatives and should be considered during new development projects (Friedman 2007).

2.2 Key Pillars of Sustainability

The three original pillars of sustainable development emerging from the WCED document *Our Common Future* were social, economic, and environmental sustainability; however, the list was amended soon after to include culture and governance as important measures of sustainability. A sustainable approach to culture and governance dictates that practices must reflect and respond to the values and traditions of the sociopolitical context in which they are implemented. While these factors have not always been considered by international efforts in the past, they have proven essential to successful development. The following section will outline key components of the five pillars of sustainable development provide essential guidance for today's planners.

2.2.1 Social

Social sustainability is often given less consideration than economic and environmental sustainability. It is important to develop social sustainability to "initiate, advocate and absorb radical shifts in desired lifestyles, values and technology" required to implement economic and environmental sustainability (Mckenzie 2004). Social sustainability incorporates the development of social capital, resilient communities, human rights, public health, labor rights, and social supports. Social sustainability is symbiotic with economic and environmental sustainability. For instance, when the creation of a sustainable healthcare system is the objective, ensuring that sufficient funds are available is essential. A contribution to public health can be achieved by encouraging fitness and installing exercise equipment in public parks (Fig. 2.2). It has been shown that people with an active lifestyle are less

Fig. 2.2 A contribution to and investment in overall public health can be achieved by encouraging fitness activities and installing exercise equipment in parks

likely to suffer from cardiovascular and diabetes-related illnesses. Therefore, it is in the best interest of municipalities that its neighborhoods are designed with bicycle and pedestrian pathways, integrating both residential and nonresidential functions. This will be explored further in Chap. 15.

2.2.2 Cultural

Cultural sustainability is equally vital. Promoting vernacular culture and preserving local traditions contributes to society in both direct and indirect ways. Old buildings are visible reminders of a society's history, creating a direct connection to the past (Fig. 2.3). Heritage sites can inspire architects and planners, giving them a view into the history of a place, thus improving the quality of future buildings and designs. Preserving and retrofitting old buildings bypasses demolition, working to reduce the consumption of natural resources that may otherwise be used in new construction.

Fig. 2.3 Cultural sustainability can be enhanced by restoring the original exteriors as was done in the Bo-Kaap neighborhood in Cape Town, South Africa

2.2.3 Economic

Economic sustainability is one of the most prioritized pillars of sustainability. A sustainable business should minimize its negative environmental impacts, consider the wellbeing of its employees and be prepared for future challenges. Ethical, sustainable business practices have become increasingly popular; public pressure has pushed businesses to invest in green technologies and develop more sustainable business practices (Pfeffer 2010). Businesses that have changed to be environmentally and socially conscious benefit from increased public support, and therefore increased revenue. In the planning of cities, current economic decisions impact the generations of tomorrow. Business practices and investments in a community will determine the opportunities that future residents will have. Building excessively wide roads, for example, will have long-term negative economic effects. In cold climates, wide streets will need to be resurfaced regularly, and a larger volume of snow removed, requiring more funding. Having shared street use by pedestrians and cars will promote walkability and also save costs (Fig. 2.4). When developments prioritize single-family dwellings the cost of service provision is significantly higher. This puts a greater strain on a home-buyer's financial security, decreasing the economic sustainability of the community.

Fig. 2.4 The introduction of shared streets reduced infrastructure cost and facilitated walkability as was the case in Sweden (left) and the Netherlands (right)

2.2.4 *Environmental*

"Environmental sustainability defines a boundary for us to satisfy our current needs without anyway compromising the quality of environment/ecosystem so that it remains equally capable of supporting the future generations too" (Kaswan et al. 2019). Sustainable development should consider the ecological impact of the construction and upkeep of a development, including its roads, open spaces, and homes. A cradle-to-cradle cycle assessment is necessary when planning a development. It is necessary to consider the initial impact of the materials used as well as their long-term performance and recyclability. Asphalt-covered roads channel rainwater runoff to manholes and into the municipal sewer system. An environmentally sustainable alternative is to create bio swells along the sides of roads to filter and collect the water, promoting the growth of native flora, minimizing input into the sewer system, and creating a more attractive built environment (Fig. 2.5).

2.2.5 *Governance*

Lastly, we must consider sustainable governance. A sustainable government should be adaptable, accountable, and have the capacity to implement long-term sustainability plans (Schraad-Tischler and Seelkopf 2015). Sustainable governance aims to develop "sustainable policy outcomes and imbue political decision-making with a longer-term focus" (Schraad-Tischler and Seelkopf 2015). Planning strategies and concepts, innovative as they may be, cannot be implemented unless a municipal leadership establishes supporting policies. Our political leaders are responsible for communicating visions of long-term planning to citizens. Further, an effective political system will encourage the participation of citizens of all ages to create a continuity of ideas and actions (Fig. 2.6).

Fig. 2.5 Having a roadside bio-swell directs rainwater back to nature like this one near Hoorn, the Netherlands

Fig. 2.6 Young participants are invited to take part in a street survey in Haarlem, the Netherlands

These five pillars critical to sustainable development can be viewed independently. Yet, when one closely examines the inner workings of cities that are designed and built using the sustainable principles, it is evident that the pillars are symbiotic. Each pillar is dependent on the others and they operate in an elaborate balance. It is this complex nexus that this book intends to clarify. While these five overlapping issues are presented separately as a way to organize concepts and illustrations, it is essential to recognize the interrelationships between them. General principles of sustainability are explored in the following section to aid in the understanding of these relationships.

2.3 The Four Principles of Sustainability in the Urban Design Context

Current methods of urban design face challenges of both philosophy and form. At the turn of the twenty-first century, urban sprawl—with its many far-reaching negative implications for society, economy, and the environment—necessitates a fundamental reconsideration and reformulation of traditional perceptions. When an attempt is made to diagnose the root cause that led to poor planning practices, ignorance of the inner workings of four pivotal issues—environment, economy, culture, and society—can arguably be one of the reasons. Mainstream developments are often regarded as a product, rather than a process, where a range of aspects are being systematically explored and manipulated. The process, the key issues and the relations among them are illustrated with four general principles (Fig. 2.7). When followed, these principles can guide the conception of a sustainable urban environment.

2.3.1 *The Path of Least Negative Impact*

The first principle is the path of least negative impact. It dictates that decision-makers must minimize any negative impact their work might have on social, cultural, economic, environmental, or governmental practices. At the outset of any

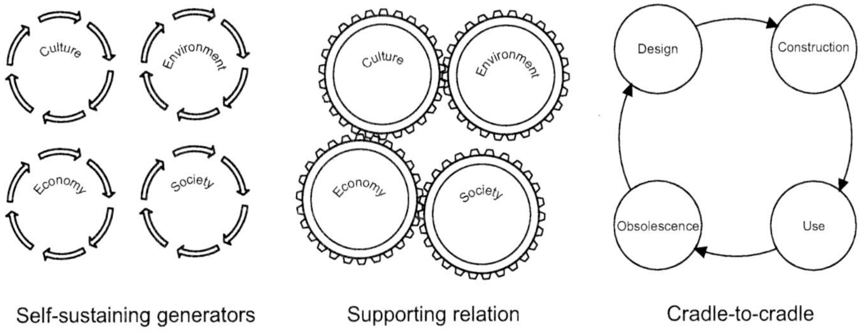

Fig. 2.7 Key principles that govern sustainable systems

Fig. 2.8 The use of permeable cover rather than asphalt as a street surfacing material, will better contribute to the returning of rainwater to nature

planning process, impact assessments should ensure limited short-term and long-term harmful and costly ramifications. An implementation of this principle is to establish an economic base that will not hinder future initiatives. Reliance on a heavily polluting industry, for example, will cause negative environmental impacts, health issues, and may limit opportunities for community development. Another example would be the use of permeable surfaces rather than asphalt as a street surfacing material which better contributes to the returning of water to nature (Fig. 2.8).

2.3.2 Self-Sustaining Systems

Establishing a self-sustaining system is the second principle of sustainability. In a fluctuating economy, reliance on a single source of municipal revenue is not sustainable. It places a community in a vulnerable position, making long-term planning difficult. Another manifestation of self-sufficiency is patronizing local businesses (Fig. 2.9). When this occurs, energy is not spent on transporting goods from faraway places, the cost of these goods is reduced, and local jobs are created. This requires, in part, microlevel behavioral change, and thus a public consciousness concerning issues of sustainability.

A self-sustaining system should be resilient, which is the capacity of a system function in the face of outside change and pressure. Community resilience is impor-

Fig. 2.9 To enhance economic activity local shopping is promoted in Woodstock, Vermont, US

tant to consider when planning. It is inevitable that events will take place in the life of a neighborhood that will demand response and adjustments for which advanced planning will be needed. Overtime communities will face new demands and given the proper skills will be able to adapt and thrive.

2.3.3 *Symbiotic Relationships*

Recognition of the symbiotic relationships between the five pillars noted above is another vital principle. The development of sustainable government, industrial, societal, cultural, and environmental programs will strengthen and support each other. For example, building homes in a dense configuration will result in a reduction of urban sprawl, improving environmental sustainability (Fig. 2.10). It will lower the cost of land, service provision, and infrastructure that, when transferred to the occupants, will result in the production of more affordable housing, improving economic sustainability. Municipalities will benefit by attracting and retaining first-time home buyers in the community and ensure a much-desired demographic continuum, improving social sustainability.

Fig. 2.10 Higher density housing reduces urban sprawl, leave more space for public parks, and contribute to the creation of affordable housing as was the case in Montreal, Quebec, Canada

2.3.4 Life Cycle Approach

The final principle to consider is a life cycle approach. This simply means adopting a series of long-term policies and objectives. It advocates that while following the course of their chosen path, a planner's overall strategy must remain flexible to accommodate ongoing changes as different needs arise including the building level (Fig. 2.11). The elasticity of any planning process and ability to adapt to various emerging circumstances are among its key attributes.

It is important to consider that when public institutions are designed for adaptability and can be easily modified to accommodate new needs, the investment required for future alterations will be smaller. A similar view should prevail, for example, when codes and bylaws are created. They ought to provide a framework for action, yet not restrict the introduction of amendments and changes when circumstances require them to. It is also essential to respect past decisions and have continuity of governance. Too often when municipal governments change, and newly elected officials are appointed, a fresh set of objectives are created, and older but sound objectives are abandoned. Continuity is essential in following a strategy with long-term implications.

Fig. 2.11 A flexible approach to façade's design simplifies adaptability of the openings to occupant's needs and personalization like in this building in Barcelona, Spain

2.4 Sustainability Indicators

In addition to using the lens of sustainability when designing urban developments, it is important to assess different measurements of sustainability; this should be done by analyzing past projects and calculating the predicted impact of current development projects. The principle indicator we can look at to assess sustainability is a carbon footprint. A carbon footprint refers to the amount of carbon dioxide emitted per person or per entity as a result of using fossil fuels. This is only one measure of environmental sustainability, but a crucial one as it is affected by all aspects of an urban development, namely transport, building materials, consumption, and waste. In general, it is a good principle to assess any indicators that are at hand; ecological indicators such as water pollution levels; health indicators such as life expectancy; and economic indicators such as employment rates, in order to determine successful practices in designing urban spaces.

The various frameworks discussed in this chapter—the aforementioned social changes of the twenty-first century, pillars of sustainable development, and principles of sustainable systems—together form a prism through which to view urban development that provides a framework and guidelines on how to move forward with sustainable development.

References

Brundtland GH, Khalid M, Agnelli S, Al-Athel S, Chidzero B (1987) Our common future. Oxford University Press, New York, p 8

Friedman A (2007) Sustainable residential development: planning and design for green neighbourhoods. McGraw-Hill, New York

Kaswan V, Choudhary M, Kumar P, Kaswan S, Bajya P (2019) Green production strategies. Elsevier Amsterdam

Mckenzie S (2004) Social sustainability: towards some definitions. Hawke Research Institute, University of South Australia, Magill, South Australia

Pfeffer J (2010) Building sustainable organizations: the human factor. Acad Manage Perspect. https://doi.org/10.5465/amp.24.1.34

Schraad-Tischler D, Seelkopf L (2015) Concept and methodology—sustainable governance indicators 2015. https://www.sgi-network.org/2019/. Accessed 6 April 2020

Part II
Urban Forms and Buildings' Shape

Chapter 3
Master Planning for Sustainability

Abstract A master plan outlines the framework of a place's present and future development. Decisions at a master planning stage will determine the use and the form of a place. Land uses, residential, commercial, industrial or institutional, will be assigned and the characteristics of structures to be built will be specified. Traffic arteries connecting these areas will be drawn. As part of a planning process, the public will be asked to participate and comment. The initiating parties may either be a land developer, a municipal or regional government and at times, a partnership of both. Once accepted and approved, it will have legal ramifications and chart all planning decisions that govern a place. The decision taken at the master planning level will also define the place's sustainability and thus it is imperative that all relevant aspects be considered.

Keywords Master plan · Land uses · Regeneration · Economic growth · Social cohesion · Environmental sustainability · Brownfield · Existing realities · High density · Mobility

3.1 Working with Built and Existing Natural Assets

A master plan is a broad term referring to the framework created for the physical development of a place. It typically consists of a top-down strategy for major change in an area, with the supporting financial, economic, and social policy documents as well as delivery mechanisms that outline its proposed implementation. The decisions made at the master planning stage are crucial in determining the use and form of a place, and in determining who will shape it. Any crowd-sourced information from local citizens will be used in the making of the master plan. The stakeholders and initiating parties are often either land developers, municipal or regional governments, or a partnership of both.

There are many ways of approaching a master plan, some of which provide more sustainable ways of conducting the process. Arguably the most important part in the making of a sustainable master plan is assessing what is already present in the area; taking stock of the existing built and natural assets and efficiently incorporating exist-

A. Friedman, *Fundamentals of Sustainable Urban Design*,
https://doi.org/10.1007/978-3-030-60865-1_3

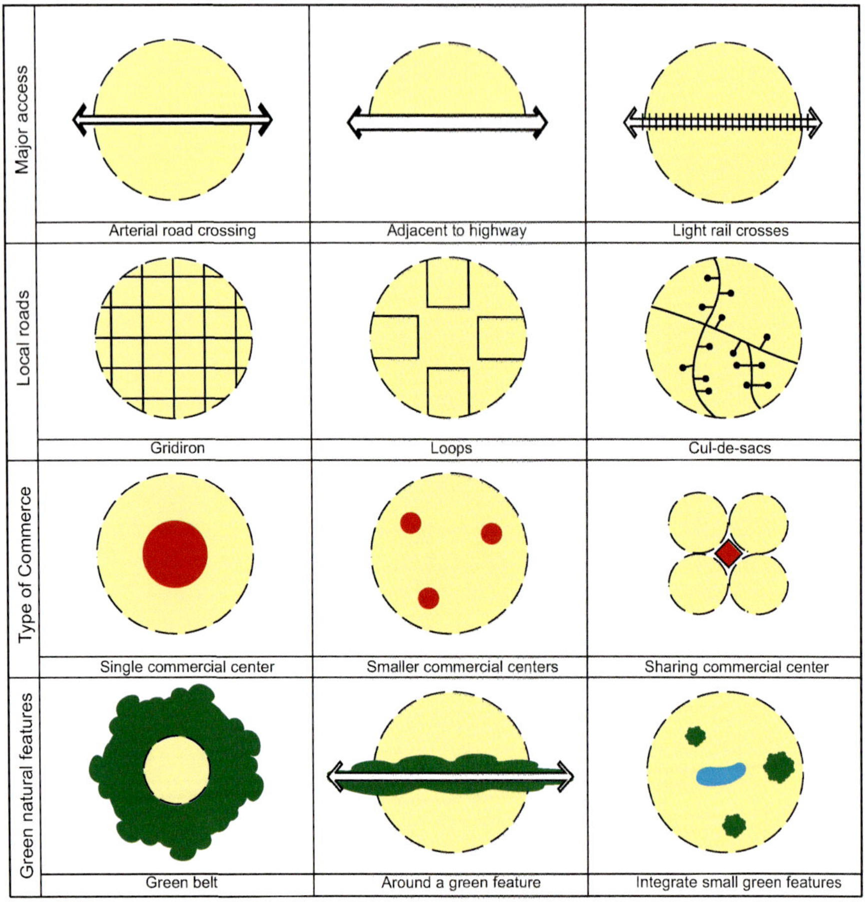

Fig. 3.1 Key elements to be considered while taking a stoke of an existing place's assets are adjacency to major access and thoroughfares, local roads, type of commerce and natural features

ing structures into proposed new developments. This can refer to a number of things, such as existing sewage infrastructure, roads and pathways, natural ecosystems or heritage buildings with historical and cultural value some of which are displayed in Fig. 3.1. It is also equally important to consider the existing social and economic realities of the area; Who lives there? What activities take place in the area? What is the history of the place? These questions merely scratch the surface regarding the research that must be conducted in order to inform the creation of the master plan.

As outlined by the United Kingdom Design Council (formerly the Commission for Architecture and the Built Environment), as well as evaluating the current context, the master plan proposes physical change in three main urban contexts: regeneration, development, and opportunities (Averly et al. 2004). Firstly, the regeneration context concerns any areas that show signs of dilapidation, disrepair, or failure. For instance, in this context, change is proposed in response to the closure of an industry resulting

Fig. 3.2 The community of Västra Hammen was developed on land vacated by the old port of Malmö, Sweden

in the availability of brownfield or redundant land for development. This is commonly the case with old steelwork areas, shipyards, docks, railway lands, or along the banks of a river and canal (Fig. 3.2). Redevelopment aimed at regeneration could also tackle areas where there is a disposal of publicly owned buildings, or where a housing estate has failed due to the quality of the housing stock and stigmatization of the community.

Secondly, the development context refers primarily to areas which previously have not been used for residential, commercial, or industrial use. Newly built towns and extensions to existing developments fall under this category. These types of developments are often created to support housing needs or stimulate investment and boost economic growth.

Lastly, under the context of opportunities, urban areas with great potential are used for a specific purpose. Developments in city centers which entail the amalgamation of sites and improvements in transport, or the creation of a major new public investment, such as new sports facilities for an international sporting event, both fall under this category of opportunity.

In each of the urban contexts outlined above, master planning is valuable in redeveloping the area, making the most of existing infrastructure and improving spaces to foster economic growth, social cohesion, and environmental sustainability, improving these areas overall (Fig. 3.3). When making plans to redevelop an area, the bigger picture must be kept in mind—and piecemeal planning must be avoided

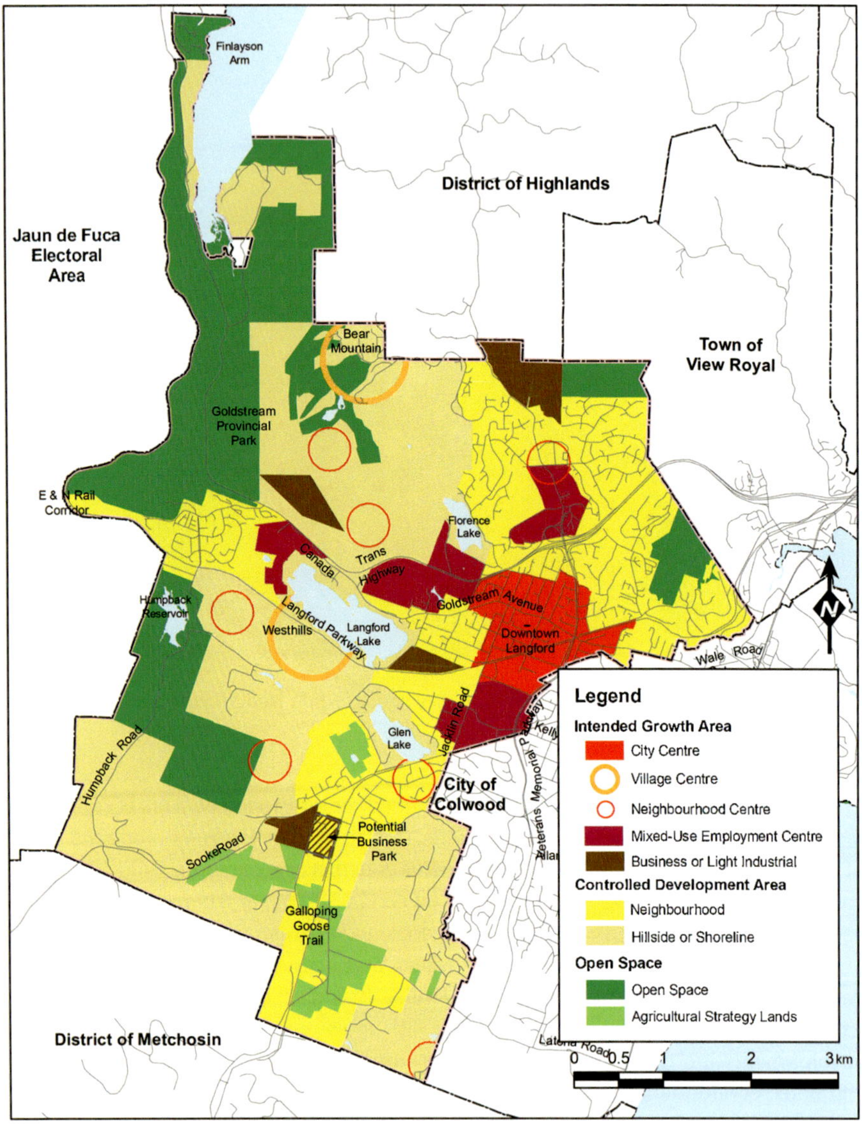

Fig. 3.3 A plan describing intended growth areas in the City of Langford, British Columbia, Canada

(Islam 2011). In his argument, Islam considers piecemeal planning to be an approach whereby a problem is solved assuming it has no impact or effect on other issues. This approach contradicts the idea of supporting relations in sustainability.

As described by David Rudlin and Shruti Hemani in their 2019 paper *Climax City*, master planning is a subtle art that works with the natural processes of urban growth, and rather than seeking to replace these natural processes within its strategy, the master plan becomes a frame into which the city can grow. This idea—of not interrupting natural processes—is echoed in the words of planning consultant

Les Sparks: "We need to avoid making everywhere like everywhere else, rather than more like itself" (Averly et al. 2004). Thus, in taking stock of existing realities, a successful master plan can be created. As outlined in this section, the process of creating a master plan can and must vary—because there is no singular master plan. Although similar approaches and strategies can be applied, a successful master plan should be unique in that it reflects the context of the place it was designed for.

3.2 Land Use Allocation and Integration for Sustainability

An important aspect of the master plan is the allocation of land use: the distribution of land to be used for residential, commercial, industrial or institutional use, or a combination of uses. The distribution of land use forms the basis of the master plan and determines the infrastructure (e.g., transportation, types of building, public space) which follows. Moreover, this stage in the process is a key stage for integrating the main pillars of sustainability (Fig. 3.4). As outlined in the Chap. 1, designing with density in mind is key in creating a sustainable neighborhood. Successful land use allocation balances transit routes, spaces, and buildings, in a manner which leaves no space unused or unoccupied: no space is left over (Averly et al. 2004). Further, building on Sect. 3.1, when allocating land use, the natural environment and topography of the area must be considered.

Good examples of land use allocation can be seen in a variety of places. For example, in the Port Whitby Sustainable Community Plan, there are several recommended land use strategies that are generally applicable to sustainable master plans (Arup et al. 2010). Namely, the promotion of mixed uses and focus on density—with residential and commercial developments concentrated around transport hubs,

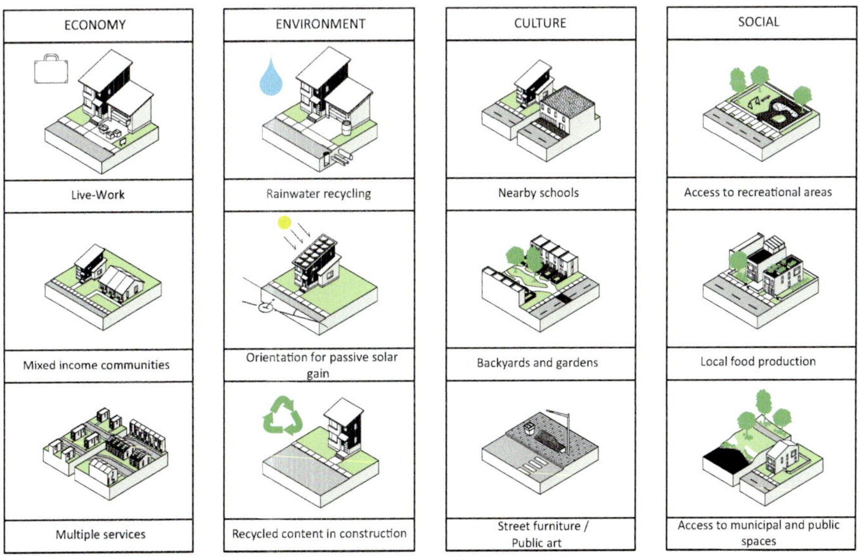

Fig. 3.4 Some key aspects to consider in the preparation of a sustainable master plan

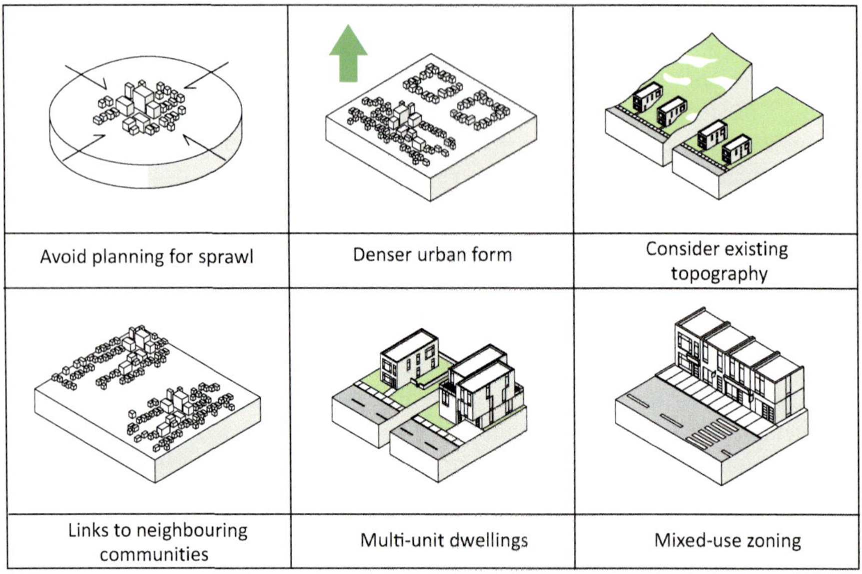

Fig. 3.5 Some key aspects to guide the planning of a sustainable community

and the provision of an active public realm, in this case a community garden space. The Gehl Master planning Frameworks provides a model of sustainable land use allocation, not only promoting mixed-use developments but also spaces which provide a wider range of activities, functions, spaces, people, building typologies, dimensions, tenures, and range of affordability (Gehl Architects 2020).

Reallocating land use in certain contexts can be sustainable in and of itself. When master plans focus on the regeneration of urban areas, land use is repurposed and put to more sustainable use. Some sustainable planning objectives are illustrated in Fig. 3.5.

3.3 Mobility and Connectivity: Reaching Amenities

With site planning comes the need to determine site layout and access, allocating not only land use but the infrastructure required to connect different areas: bike lanes, pedestrian paths, roads, and railways. Successful master plans create spaces which are well connected and often target areas that lack in these linkages, such as brownfield regeneration sites, which have often remained under-developed precisely because of lack of connectivity (Averly et al. 2004). It is important to consider the access between areas with different types of land use: a successful sustainable master plan aims to make all areas accessible. The illustration below depicts the existing and proposed road network that was proposed by the author for the Canadian town of Stony Plain (Fig. 3.6).

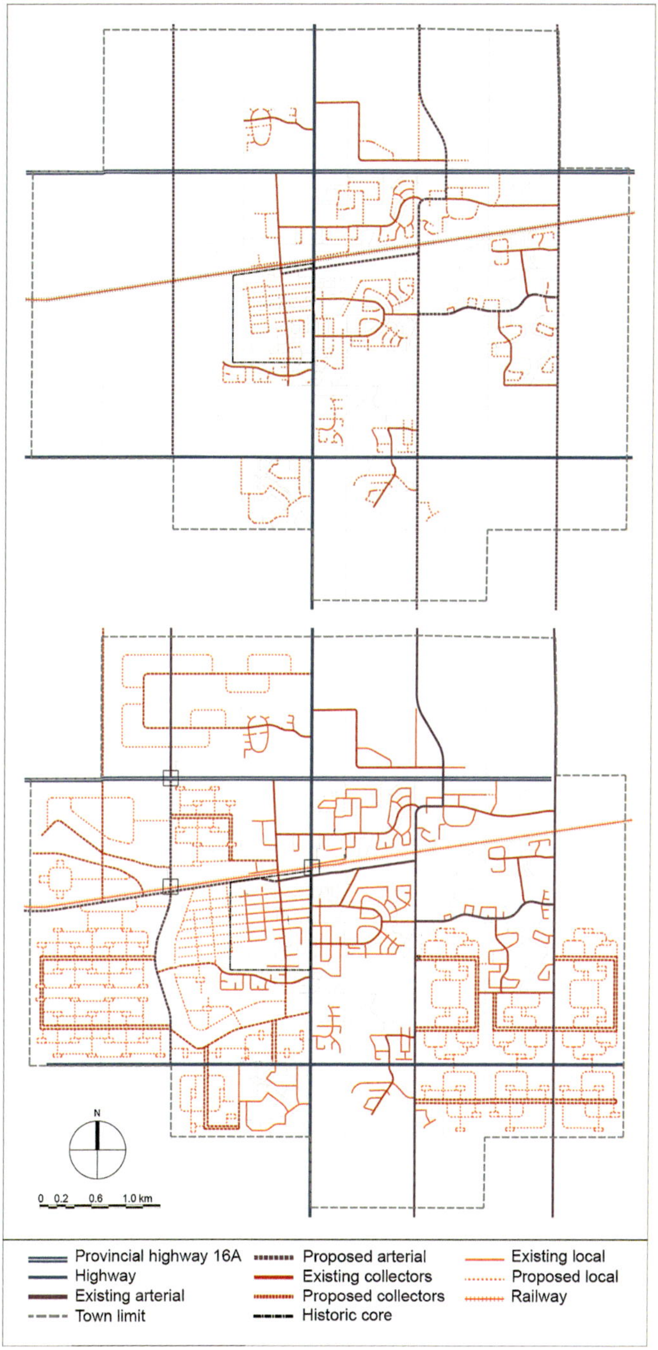

Fig. 3.6 Existing and proposed roads' network in the Town of Stony Plain, Alberta, Canada

Increased mobility goes together with planning for high density by reducing the distance between amenities and neighborhoods, as discussed in Chap. 1. It is crucial to account for mobility in the initial stages of the master plan because high mobility is dependent on the infrastructure designed to support it. Redeveloping an area also provides the unique opportunity to renovate its mobility structure, such as increase the number of bike lanes or pedestrian walkways. Once again, Gehl Architects offers a number of exemplary cases of creating master plans to promote mobility, namely, the redevelopment of Times Square in New York City, where 0.4 km^2 (0.15 sq. miles) of space was reclaimed from traffic for pedestrians. As a result, residents and tourists enjoy Times Square more—sit down, socialize, people watch—with 86% more people stopping there, and there was a 36% decrease in pedestrian and cyclist injuries (Gehl Architects 2014).

As outlined in this chapter, when considering redevelopment in an area, the initial step is assessing the existing realities in that space and determining what the space is used for and what needs to change. Tackling a master plan from a sustainable perspective means acknowledging existing infrastructure and physical aspects of a place and incorporating these in a new design—not starting from scratch. Creating a master plan provides the opportunity to apply what is known about creating sustainable and livable neighborhoods by renovating spaces with the principles of high density, mixed land use, and mobility in mind.

References

Arup, Meridian Planning, Planning Alliance (2010) Port Whitby sustainable community plan. https://www.whitby.ca/en/townhall/portwhitbysustainablecommunityplan.asp. Accessed 6 Feb 2020

Averly J, Eley J, Stewart P, Nicolaou L (2004) Creating successful masterplans. Commission for Architecture and the Built Environment, London

Gehl Architects (2014) World class streets: remaking new York City's public realm. Gehl Architects, New York

Gehl Architects (2020) Masterplanning frameworks. Gehl Architects, New York

Islam S (2011) Traditional urban planning approaches and sustainable city. Openhouse Int 36(2):15–23

Rudlin D, Hemani S (2019) Climax City: masterplanning and the complexity of urban growth. RIBA Publishing, London

Chapter 4
The Form of a Place

Abstract The form of a place as governed by its shape, natural assets, original planning, and evolution will dominate its sustainability potential to a large measure. For example, when a place was initially designed in a concentric way compared to an elongated one, the distance between its center and the edges will be smaller, leading to a shorter travel time. Similarly, when a community was master planned to evolve around a center in an organic fashion other advantages will be noticeable. This chapter recalls the main planning forms and their effect on urban design of communities.

Keywords Natural assets · Human settlements · Ecological footprint · Spontaneous settlements · Planned cities · Grid layout · Ebenezer Howard · Garden cities · Letchworth · Radburn · Euclidean zoning · New urbanism · Smart growth

4.1 Historic Evolution and Typology of Urban Places' Forms

A variety of research has attempted to explain the origin of human settlements, and of these investigations, the work of Jane Jacobs is an excellent starting point. In the *Economy of Cities* (1969), Jacobs suggests that the origin of human settlement is rooted in their transition to agriculture and trade. Prior to that, most societies were predominantly nomadic and lived in bands of hunter-gatherers, surviving on what nature could provide. In these groups, population size and ecological footprint were determined by food supply (Schoenauer 2000). This way of life was transformed with the invention of agriculture, meaning food availability no longer dictated a group's size.

Historian Spiro Kostof (1991) argues that the interest of the authority, rather than any particular form of activity, led to the establishment of many towns. Governments—both centrally organized and distant seats of power—settled and populated early communities of loyal subjects for political gain as a means to establish and expand their power. For example, many European cities grew from their Roman origins.

A. Friedman, *Fundamentals of Sustainable Urban Design*,
https://doi.org/10.1007/978-3-030-60865-1_4

Fig. 4.1 According to Kostof, settlements, be they cities or neighborhoods, follow two distinct patterns of urban form. The first were planned or created cities like Kamiros, Greece (top) and the second were spontaneous like Volterra, Italy (bottom)

Kostof also suggests that communities, be they cities or neighborhoods, follow two distinct patterns of urban form. The first are planned cities (also known as created cities); these are places whose form follows a charted plan much like post-Second World War suburban towns. The second kind is that of spontaneous settlements, communities that grow by a less regulated process (Fig. 4.1).

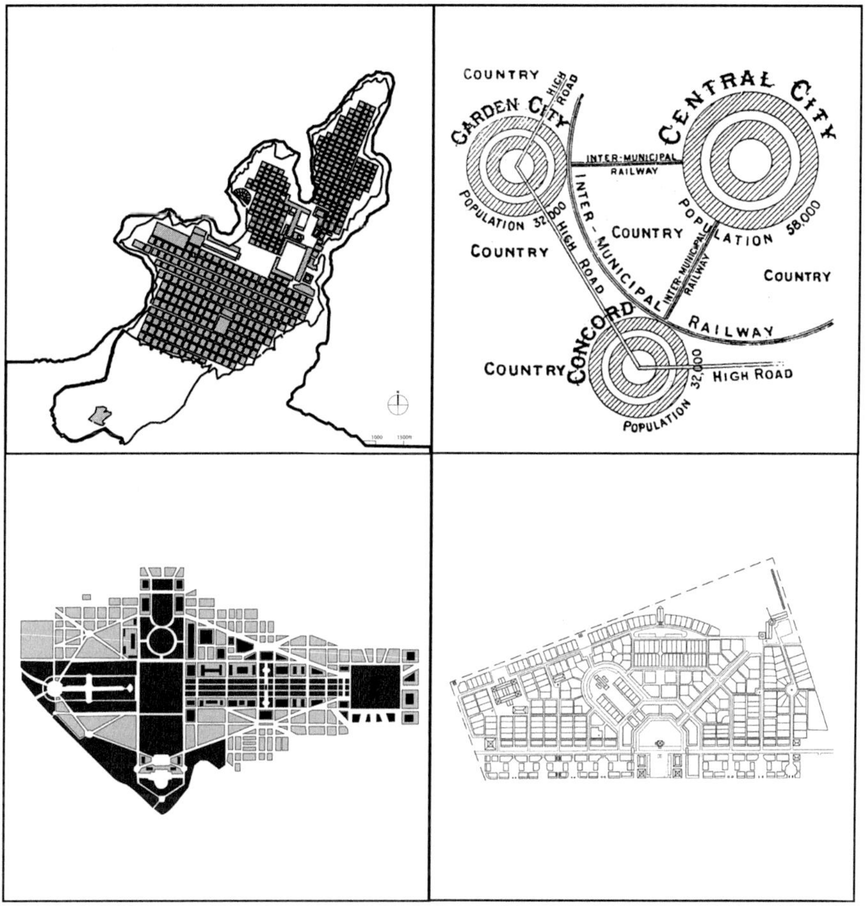

Fig. 4.2 Some urban planning typologies for cities and neighborhoods: Grid layout in the ancient Greek city of Miletus (top left), Howard's diagram of Garden City (top right), Grand Manner used in the planning of Washington, DC (bottom left) and New Urbanism in Seaside, Florida (bottom right)

Kostof goes on to identify four urban planning typologies that cities and neighborhoods are likely to follow, each with a different appearance and function (Fig. 4.2). The first layout that was developed by way of uncontrolled growth is categorized "organic"; many medieval cities fit this pattern. Kostof's second type, a "grid layout," is based on geometrical or orthogonal principles that are thought to have been developed by the Greeks and spread far afield by the Romans. Rooted in the Milesian, named after the town of Miletus, neighborhoods of this form are divided into relatively autonomous areas. Further, there is the "diagram pattern" based on formal geometry. Kostof dictated that cities of this type are designed by an individual in accordance with their ideals—cities are manifestations of a vision as to how a community should function and how its inhabitants should live. Ben-Joseph and Gordon (2000) suggest that such geometrical schemes were rarely built, condemned to a life only on paper. The final category of Kostof's typology is that of the "grand manner." In these designs, buildings, streets, and public spaces are arranged to convey visual effect of grandeur and coherence. Such patterns are typical of the Renaissance.

Moving forward, contemporary planning practices can in good part be traced back to the thinking of Ebenezer Howard. In late nineteenth century England, Howard was exposed to the ongoing debates surrounding the squalid conditions suffered by urban residents. In response, he conceived of a new type of community—one that attempted to draw upon the best that both city and country living could offer. This led to his seminal work in 1898, *To-Morrow: A Peaceful Path to Real Reform* (later renamed *Garden Cities of Tomorrow).* In it, he argued for the population to withdraw from overcrowded industrialized cities to their outskirts, thus creating communities that would combine the social conveniences of towns with the healthier, more peaceful aspects of rural life.

Howard's *Garden Cities of Tomorrow* deserves close consideration. His original proposal included a city population cap of 32,000, which was to ensure that the benefits of rural life were protected. The design itself is composed of layered circles, but it does not specify a particular architectural style for the buildings involved. Industry is centrally located, surrounded by a ring of parkland. Around this park is a crystal palace: a glass arcade that houses a shopping area. The next layers were to be houses with attached gardens (Howard 1902). Howard divided his diagram city into six identical wards, each housing 5000 people; he proposed locating schools in the heart of each ward, in an open space designated Grand Avenue (Fig. 4.3).

It was the first time that the ideal population for a neighborhood had been precisely delineated in modern times. This was not the only standard that Howard set. Many of the more original components of the work became dominant elements in designing contemporary communities: defined population size, hierarchical order, inclusion of green spaces, a public transit link to the big city, and land-use distribution between industry, agriculture, and housing (Friedman 2017).

Howard's plans were realized, an obvious further measure of influence. Most notably, they were followed quite closely in the design of the town of Letchworth. In 1903, the Garden City Pioneer Company bought 1529 hectares (3822 acres) of land north of London, in the UK. Raymond Unwin and Barry Parker were hired to translate Howard's diagram for the site. A centralized civic area in Letchworth was enclosed by a park, and housing radiated from this center.

Years later, the town of Hampstead Garden Suburb was planned and built. Designed as well by Unwin and Parker, it included some of Howard's principles (Miller 1992). The next stage in the evolution of the Garden City schema was to cross the Atlantic, where it was to take on a new life in the American context.

In 1922, two American planners, Clarence Stein and Henry Wright, traveled to England to visit Letchworth. Upon returning to the USA, they informed a small group of colleagues of what they had learned. The group, which came to be known as the Regional Planning Association, discussed and incorporated many avant-garde planning and social ideas of the time. Their goal was to rethink the planning of

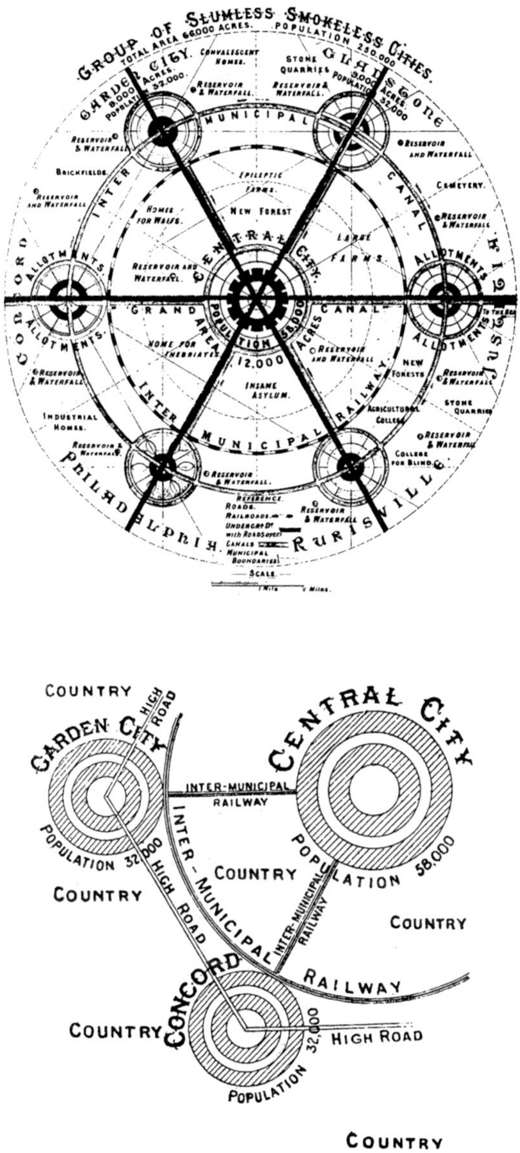

Fig. 4.3 In his plan for a *Garden City*, Howard divided his diagram into six identical wards, each housing 5000 people; in the heart of each ward, in an open space designated as Grand Avenue, he proposed a school

America's large, industrial cities—then beset with destitute living conditions—and replace them with more human environments. They sought to reach these goals primarily by following the innovations in the Garden City. Stein and Wright went on to design communities that would embody that vision. Radburn, New Jersey, USA, is the most renowned product of their partnership. Houses were sold in Radburn as opposed to being rented in garden cities and new towns in the UK. Though elements were sacrificed in this implementation (for example, the greenbelt surrounding the

Fig. 4.4 The master plan of Radburn, New Jersey included a variety of housing types, and neighborhoods with cul-de-sacs and scenic, curving streets

town was never purchased because of the Great Depression), the overall result was a safe, healthy community for young families. Radburn also included a variety of housing types, and neighborhoods were serviced by small retail centers. Their forms included cul-de-sacs and scenic, curving streets (Fig. 4.4).

The success of Radburn was due in part to its accommodation of the automobile whereby the pedestrian and the automobile were completely separated. This new phenomenon stimulated many novel development patterns. Circulation designs were used to separate pedestrian and vehicular traffic via interior paths and overpasses. Although most dwellings in Radburn were single-family, some units existed in garden apartments (Schoenauer 2000). Individual unit planning was oriented toward the internal open areas rather than to the streets. These design articulations made Radburn a model of planning in suburbia for the following decades, notably the suburbs built post-Second World War. But perhaps the greatest impact on suburban planning that emerged was that of segregated land-use planning.

Euclidean zoning, which separated large tracts of uniquely residential areas from all other occupancies, emerged in 1926. Zoning is the separation of land use, isolating residential, commercial, and industrial functions of a city (this will be explored more in Chap. 9). It was created to simplify the speculative development process and also resulted in geographical segregation by income (Logan 1976). Member of the Regional Planning Association drew inspiration from the most influential intellectuals of their day, such as Clarence Arthur Perry, a planner, sociologist, author, and educator.

Similar to Howard, the driving forces behind Perry's pivotal ideals were social needs. In 1907, he went on to work for the Russell Sage Foundation that was established to improve living conditions in cities. He resided in Forest Hills Gardens, a Sage Foundation development, in Queens, New York, USA. Planned by Frederick Law Olmstead Jr., this experience helped him define his neighborhood unit concept.

When observing Forest Hill, Perry noted five aspects that contributed to its success: clear boundaries, an internal street system of superior design, a variety of well-chosen land uses, the provision of open spaces, and, vitally, the presence of a central area (Fig. 4.5). Based on his observations, he went on to conceive of his own innovative planning principles. First, Perry designated that the unit needed to have a shape wherein all sides were of equal distance from the center. He suggested an ideal 0.4 km (0.25 mile) radius and a size of 65 hectares (160 acres). The neighborhood center was to have communal amenities including a school in a central green space, while shops were to be located in outer corners. Small parks and open spaces were to be scattered in each quadrant, and account for 10% of the neighborhood area. Arterial roads were to be bound to the unit's sides. Finally, Perry dictated that streets were to be both curvilinear and straight to reduce traffic.

Perry's contribution to modern planning was not only in the simplicity and logic of the plans themselves; like Howard, his utilization of dimensions and figures that planners could easily grasp set an important precedent. Another aspect of his influence, not to be underestimated, was that his plan ushered neighborhood planning out of the carriage era and into the automobile age.

Howard, Perry, Stein, and Wright are prominent men in the effort to improve the standard of living in the twentieth century. Their ideas went on to offer dwellers of crowded cities an alternative by inventing a new form of community living; each family enjoyed a house and a yard of their own. Yet, these benefits were not to be without cost; rampant consumerism, long commutes, social segregation, social isolation, high pollution, and the exhaustion of natural resources have become commonplace by-products of this lifestyle.

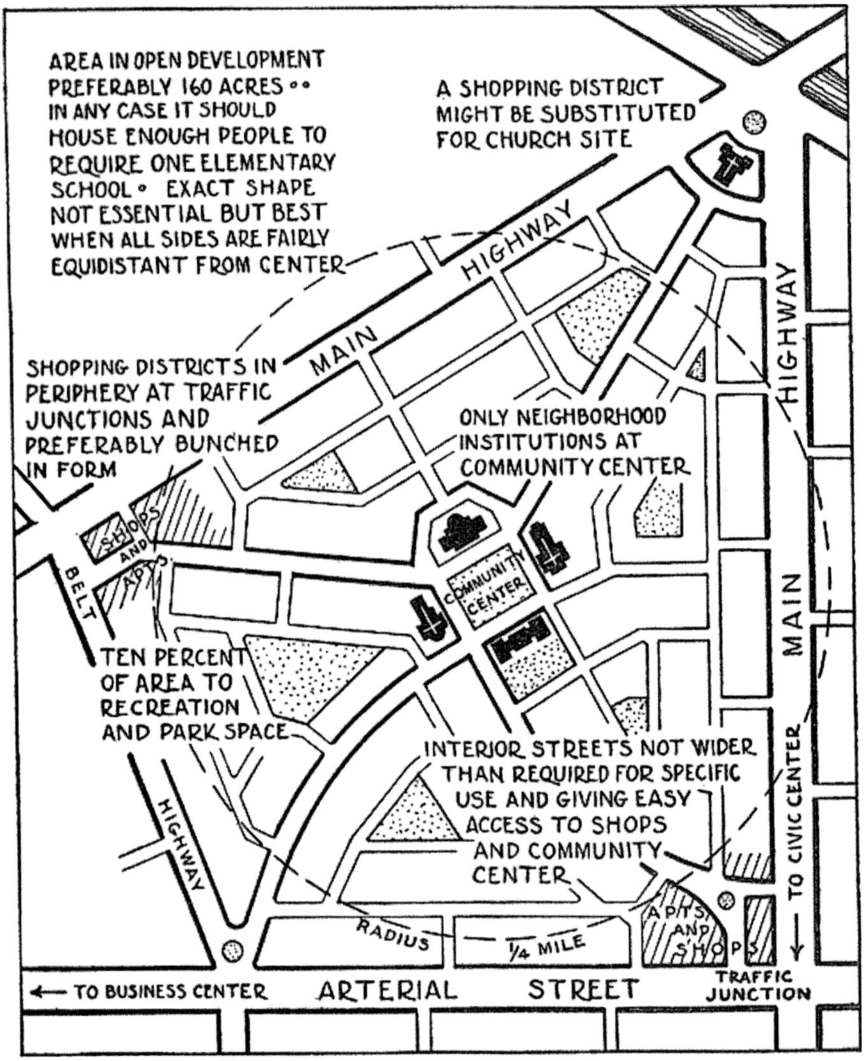

Fig. 4.5 For his "Neighborhood Unit" Clarence Perry listed five aspects that contribute to a neighborhood's successes: clear boundaries, an internal street system of superior design, a variety of well-chosen land uses, the provision of open spaces, and, vitally, the presence of a central community area

4.2 The Form of a Place and Sustainability

The previous section of this chapter outlined the evolution of urban planning. Suburban development resulted in multiple harmful effects on the environment and society. These were due to decisions made in the planning process. For instance, roads and pedestrian walkways were separated and vehicles were allocated the greatest amount of space, encouraging vehicle dependence in the system. Another example is that of residential lease agreements which determined who could purchase property and in which area, leading to racial and social segregation (Freixas and Abbott 2018).

While the negative consequences of suburban development are clear, it was a significant improvement to earlier settlements which were prone to overcrowding, disease, and pollution. Modern planners must separate the positive design features, such as incorporating green space, from the negative ones, such as vehicle dependence. In more recent times, planning concepts such as "New Urbanism" and "Smart Growth" attempted to offer an alternative to suburban sprawl. Put forward by Andres Duany, "transect" defines zones that range from rural areas to urban centers with each having scales that link developments to one another with a mix of residential and commercial land uses. Designing urban developments around a center with common commercial and public space allows for greater access and proximity to useful services, allowing residents to enjoy the convenience of colocation (Sim 2019).

In relation to land and urban form, a sustainable urban development incorporates one key concept: decrease land use. Firstly, at the initial stages of urban development, the conversion of valuable natural environment should be avoided and minimized. Secondly, development should be compact, consisting of higher density settlements which consume less land (see Chap. 5) (Fig. 4.6) (Pradhan 2017). In essence, the aim is to follow the path of least negative impact that was discussed in Chap. 2. The goal is to minimize environmental harm caused by humans occupying the land. Moreover, allowing for a more compact urban form allows for a plethora of positive effects such as: shorter travel distances and times, decrease reliance on cars, better public transportation services, increased active mobility, increased overall accessibility, regeneration of existing urban areas and urban vitality, preservation of green spaces, reduced energy consumption, and overall a higher quality of life (Pradhan 2017).

Limiting outward expansion and strengthening the core was the objective in the retooling of the town of Stony Plain in Alberta, Canada by the author and his team. Located 33 km (20 miles) west of Edmonton, the town's population of 15,000 people is spread over 17 km^2 (6.56 sq. miles). The town is linked via two highways, which cross it, to the capital region.

To chart Stony Plain's future urban development along sustainable principles, we considered strategies that strengthen its economic performance, environmental conditions, social profile, and cultural attributes. On the foundation of the existing land uses we reformulated the town's master plan. In the new proposal, illustrated in Fig. 4.7, we preserved some of the former land uses and adjusted residential, commercial, and industrial areas. The new plan proposes to place less dense residential area on the town's north-west and denser neighborhoods south of the highway.

To strengthen the core, we recommended that the town limit the building of large stores south of the highway and encourage the construction of taller apartment buildings with ground floor businesses.

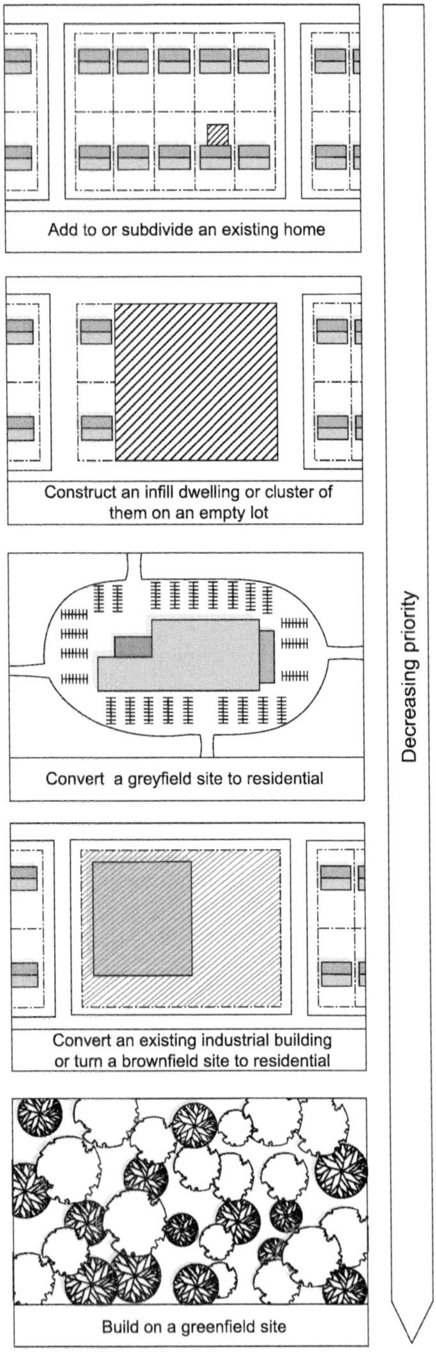

Fig. 4.6 Planning for sustainability seeks to develop a project in an existing built area first rather than build on a green field

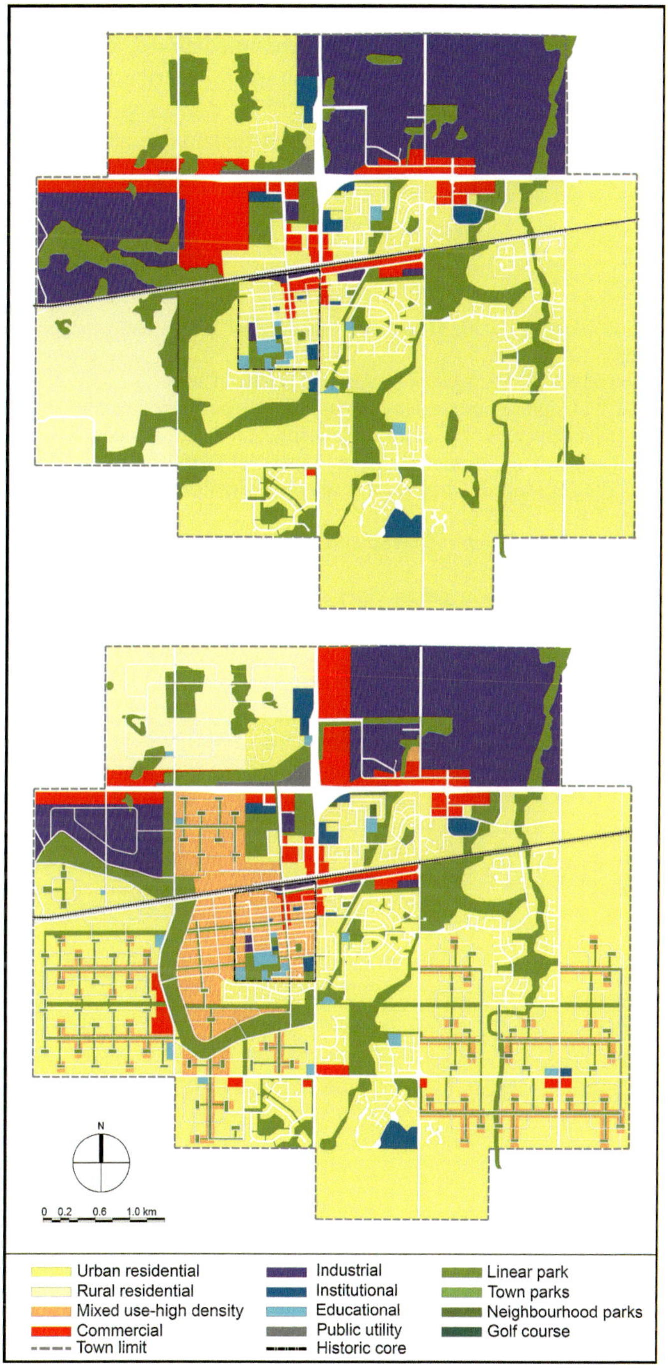

Fig. 4.7 The new master plan for the Town of Stony Plain, in Alberta, Canada proposes to place less dense residential area on the town's north-west and build denser neighborhoods around the old center

Compact cities are widely considered the most sustainable form of urban development, due to the key characteristics that were noted above, and are related to the shape and pattern of urban features, such as the spatial distribution, land use categories, and the pattern of road networks. As will be discussed below, planning for increased density is imperative to the creation of sustainable urban plans.

References

Ben-Joseph E, Gordon D (2000) Hexagonal planning in theory and practice. J Urban Des 5(3):237–265
Freixas C, Abbott M (2018) Segregation by design. Springer, Cham
Friedman A (2017) Designing sustainable communities. Bloomsbury, London
Howard E (1902) Garden cities of to-morrow. Sonnenschein & Co., London
Jacobs J (1969) The economy of cities. Random House, New York
Kostof S (1991) The city shaped: urban patterns and meanings through history. Thames & Hudson, London
Logan JR (1976) Industrialization and the stratification of cities in suburban regions. Am J Sociol 82(2):333–348
Miller M (1992) Raymond Unwin: garden cities and town planning. Burns & Oates, London
Pradhan B (2017) Spatial modelling and assessment of urban form. Springer, Cham
Schoenauer N (2000) 6,000 years of housing. WW Norton, New York
Sim D (2019) Soft city: building density for everyday life. Island Press, Washington, DC

Chapter 5
Choosing a Suitable Density

Abstract The density of a place will have a significant impact on its sustainability. It will determine its carbon and the ecological footprint, among others. Very low density will not justify financial investment in public transit, local commerce, and amenities. On the other hand, higher density is known to save energy since less people drive, choosing to walk instead, contributing to their health. Density will also affect urban design; sparsely populated places with low-rise buildings will have a different appearance than their city's counterpart. This chapter sets indexes for a sustainable density and describes its ramifications for urban design.

Keywords Public transit · Local commerce · Sustainable density · Residential density · Net density · Gross density · Floor area ratio · Pedestrian network · Open spaces · High density · Mixed densities

5.1 Principle Density Indices

The density of a community will determine many characteristics of the urban landscape. Not only is density a key feature of appearance, it determines a multitude of planning considerations including zoning and housing affordability. In the realm of environmental sustainability, density can also have great effects; relating directly to how we accommodate and build our infrastructure and use the natural resources at hand. The measurements of density used as indices of community planning and appearance are elaborated in the following section.

First, "residential density" commonly refers to the number of inhabitants or dwellings per unit area (Friedman 2012). Second, "net density" refers to the number of dwelling units divided by the total area of the lots allocated for housing, which includes land occupied by front, back and side yards, driveways, and garages. Third, "gross density" refers to the number of dwelling units divided by the area of the lots plus one-half the area of bounding streets and one-quarter the area of bounding streets intersections. This form of density, which is always lower than net density, identifies the relationship between the built mass and the open space, providing a good indication of the amount of open, often public areas such as parks and a good

A. Friedman, *Fundamentals of Sustainable Urban Design*,
https://doi.org/10.1007/978-3-030-60865-1_5

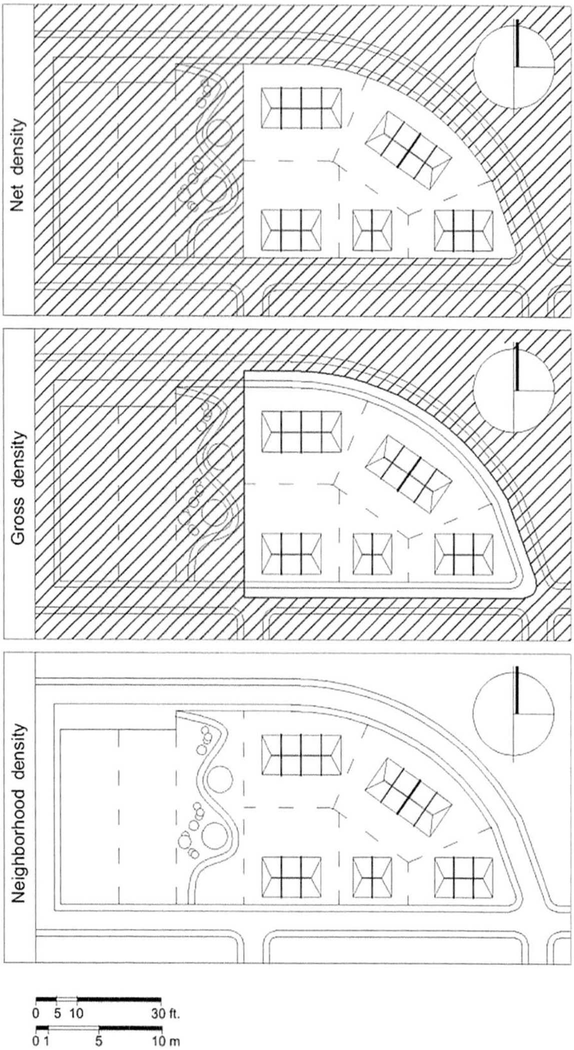

Fig. 5.1 Various density indices that help gauge the community's urban character. The shaded area is excluded from calculations

indication of the quality of life the residents will have. Fourth, "neighborhood density" divides the number of dwellings by the total area, including land used for dwellings, streets, recreation, public institutions, and shopping (Fig. 5.1). The latter gives a good sense of the general urban character of the area. Finally, another useful index for determining the density of a development is the Floor Area Ratio (FAR), which is calculated by dividing the total habitable floor area of all stories of a house by the total area of the lot. Since it establishes a relationship between land area, building floor area, and height, FAR is considered one of the most accurate measures of light and air quality of a community (Pantoja 1983). Each measurement of density has its own advantages for analyzing the spread of buildings and infrastructure in a defined area, and all the data collected can be assessed to improve the economic and environmental performance of the urban landscape.

5.2 Densities for a Sustainable Development

Projects can be categorized as low, medium, or high density, each of which has unique advantages and disadvantages. As illustrated in Fig. 5.2, low-density projects consume large amounts of land, resulting in increased infrastructure and road costs. Neighborhood amenities can be difficult to reach and often involve taking a car, particularly in cases where installing public transit networks is not economically viable. Consequently, due to a high reliance on private vehicles, there are significant negative environmental consequences. That said, residents of low-density communities have unrivalled privacy. In medium density projects, a pedestrian network can be introduced because amenities are located in closer proximity, which also allows for a feasible transit network. Moreover, infrastructure costs are reduced, and some land preserved as public space can become an extension of private outdoor areas. Density at its highest form can lead to issues of overcrowding and moreover become dense beyond the human scale. Thus, a balance needs to be found between high density and livability.

There is a strong case to be made for achieving a high, yet livable density. As argued by Sim, with rapid urbanization and scarce resources, we have to use existing infrastructure more efficient and make better use of the resources we have—specifically the space we have (Sim 2019). There is no doubt that using this approach and minimizing the environmental impact of urban developments by using land in a more efficient way is the sustainable option. A sensible increase in density and strategic planning will ultimately result in reduced cost of land, infrastructure, public roads, building materials, and energy which increases sustainability and leads to more attractive built environments (Fig. 5.3). Moreover, according to Bramley et al. (2009), higher densities may make access to key services and facilities more economically viable.

Furthermore, high density is preferable from a social point of view: Gehl argues that if activities and people are assembled, it is possible for individual events to stimulate one another (Gehl 2011). In other words, if people are living closer to one another and gathered in the same area for different purposes, more social interaction occurs and the quality of urban life increases. This also relates to Sim's equation of

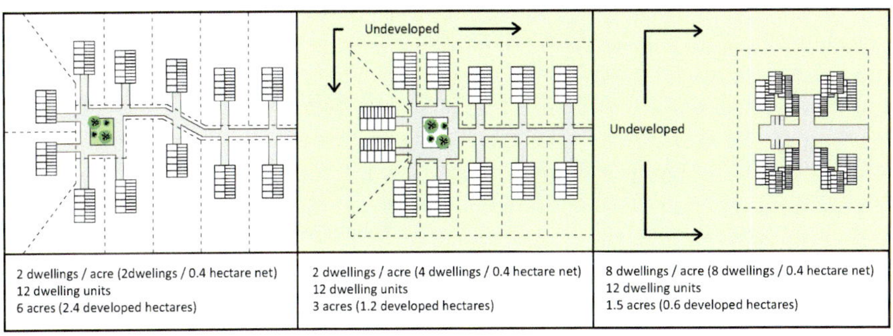

Fig. 5.2 Low-density projects consume large amounts of land, resulting in increased infrastructure and road costs

Fig. 5.3 A sensible increase in density and proper planning will ultimately result in reduced cost of land, infrastructure, public roads, building materials, and attractive built environments

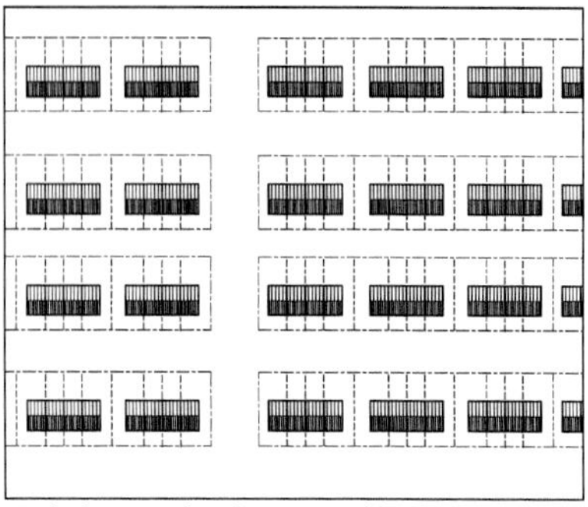

High density developments like this 31 units per acre (77.5 units per hectare) form of the 20th century is likely to be unwelcome by many suburban towns and unfavourable by buyers.

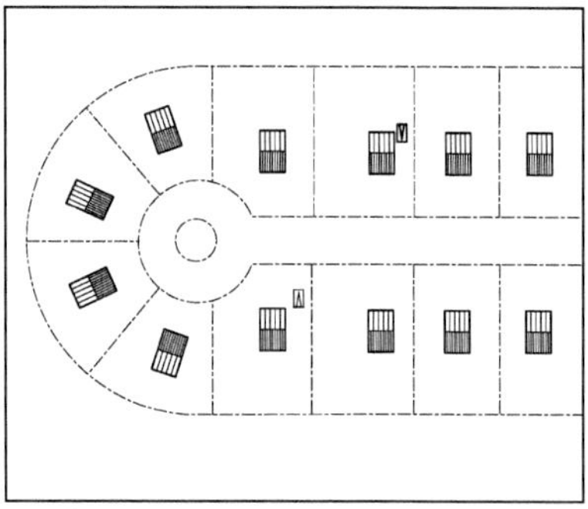

Dwellings with a density of 7 units per acre (17 units per hectare) are common in contemporary suburbs.

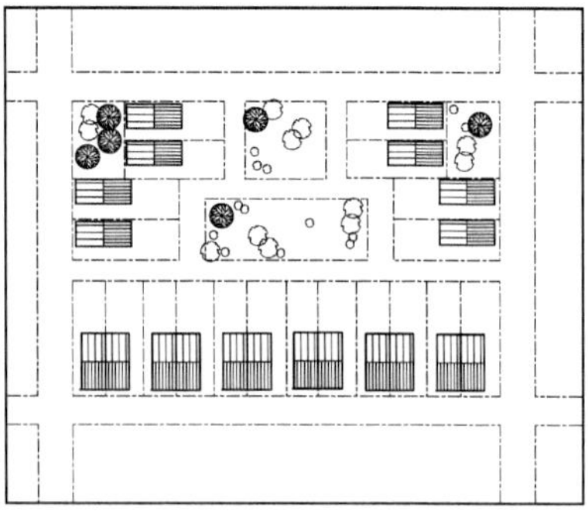

A community with a medium density of 20 to 25 units per acre (50 to 63 units per hectare) can leave open space for residents to enjoy when properly planned.

density and diversity giving proximity; the idea that the fusion of density and diversity increases the likelihood or the possibility of useful things, places, and people being closer to you (Sim 2019).

5.3 Targeting Density to Fit Urban Forms

The appropriate density for any development will largely depend on the context in which it is proposed, and in many cases, the planning of an urban development can mediate between low and high density (Friedman 2014). As shown in Fig. 5.4, the plan may designate areas to high density that would economically justify introduction of commerce and public transit for example. With this model, more people live in close proximity to transit hubs which link to other urban centers, incentivizing the use of public transit and promoting transit-oriented develop-

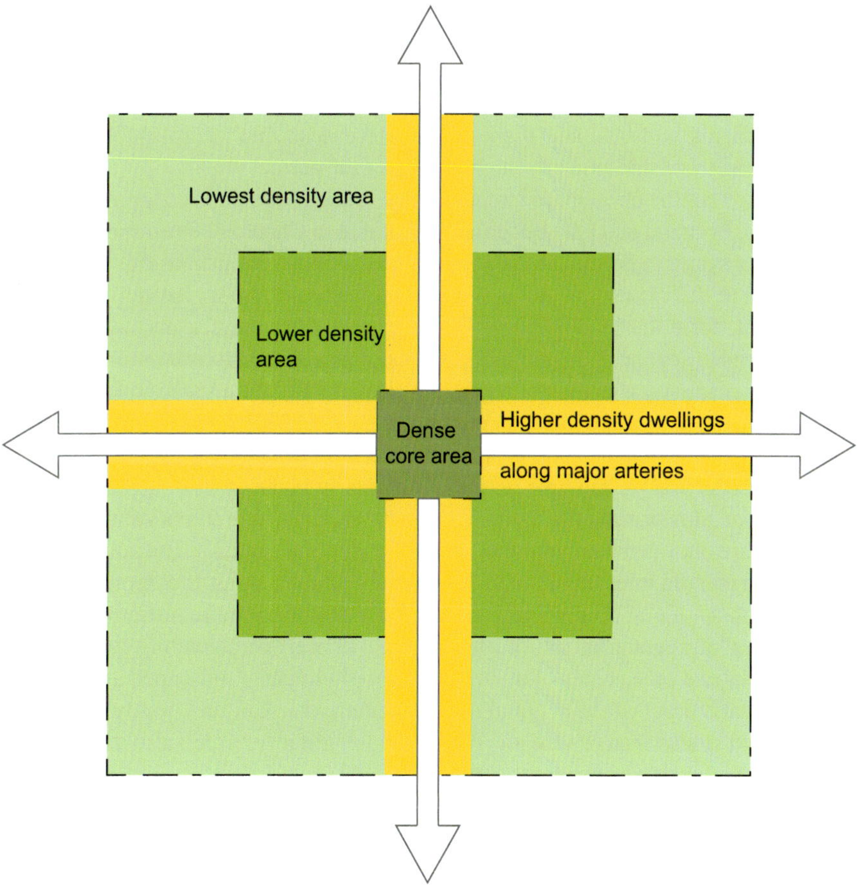

Fig. 5.4 City planning can mediate between low and high densities. The plan may designate areas to high density that would economically justify introduction of commerce and public transit

ment. Moreover, public spaces and buildings should also be located near the core area, since it helps create a strong sense of community, comfort, and safety within the neighborhood. On the periphery, the density can be gradually reduced to have more single-family homes. Lower density housing can be sited to make walking to a commercial hub no longer than 10–15 min for example. Thus, different densities can be mixed and used in conjunction around the existing factors determining urban form, such as arterial roads, topography of the area, existing bodies of water, and transportation links.

5.4 Balancing Higher Densities with Open Spaces

Public outdoor spaces are integral to high quality urban living as they are used for recreation, social integration, and physical activity. Retreats, such as parks of various sizes, should be designed in a multipurpose fashion and appeal to young and old members of the community (Fig. 5.5). They should be well-distributed, be visible and accessible. Accordingly, it is preferable to have open public spaces in built-up areas, accommodating any available space be it between buildings, in laneways or within residential developments. Creating green spaces and incorporating existing natural elements within a site makes for a more desirable living space and thus should be a key consideration in the planning of the area.

Deciding what the open space system will be is another stage in conceiving a high-density neighborhood and will be further elaborated in Chap. 24. Open spaces can be regarded as a system where an easily accessible network of green areas that range from the regional to individual unit levels is created. Large scale parks, located outside the development, form the most public of these spaces. Alternatively, enclosed outdoor areas are private spaces. Between these two extremes are semiprivate spaces such as community gardens and communal areas for clusters of homes; spaces open to members of a certain neighborhood. As density increases, the importance of open spaces also rises since the amount of space allocated to each home declines. These open spaces are largely beneficial for multiple reasons. Particularly in the cases of parks and green spaces, which allow natural life to thrive (Sim 2019), but also in forms such as a courtyard or square, they provide important spaces for informal social interaction.

Dense preserved trees can also be designated as public areas. Incorporating these in the city plan—along boulevards for example—can bring the acoustic and absorbing qualities of vegetation into built-up areas, as well as providing an aesthetic appeal. Parks and other types of open spaces should inhabit landscaped areas which have been carved out of the original natural system: the parkland, for example, can continue and spread into private lots to maintain continuity, which also makes these areas more accessible to residents (Friedman 2012).

For example, Singapore serves as an exemplary model of successfully incorporating green spaces in a densely built urban environment, namely through the Park Connector Network which links different natural parks within the city (Kaw et al. 2020).

Fig. 5.5 Parks of various sizes should be designed in multipurpose fashion that appeal to young and old people

Smaller pockets of open space are more appropriate in higher density planning as they are safer and less intimidating as opposed to less private vast open spaces (Fig. 5.6). Large open cleared areas can be separated from the more serene areas by flora. For an open space bounded by buildings, there needs to be adequate roads or pedestrian paths to encourage its use. Access points can be readily visible and accommodating, creating a sense of public place and furthermore, infrastructure

Fig. 5.6 Smaller public open spaces, also known as Pocket Parks, are more appropriate in higher density planning as they are safer and less intimidating as opposed to less private vast open spaces

such as benches, playgrounds, and seating areas is essential in providing an incentive to stay and use these green spaces.

Density is a key factor in achieving sustainable urban development. In general, the denser a neighborhood is, the more sustainable it is as it poses less harm to the environment and creates spaces which foster economic and social sustainability. That said, urban developments can also achieve sustainability by combining a range of densities and incorporating open spaces in accordance with existing urban form, in order to accommodate the various needs of the development.

References

Bramley G, Dempsey N, Power S, Brown C, Watkins D (2009) Social sustainability and urban form: evidence from five British cities. Environ Plann A 41(9):2125–2142

Friedman A (2012) Town and terraced housing: for affordability and sustainability. Routledge, New York

Friedman A (2014) Planning small and mid-sized towns: designing and retrofitting for sustainability. Routledge, New York

Gehl J (2011) Life between buildings. Island Press, Washington, DC

Kaw JK, Lee H, Wahba S (2020) The hidden wealth of cities: creating, financing and managing public spaces. The World Bank Group, Washington, DC

Pantoja AH (1983) Site planning for low-rise housing, with special reference to northern climates. Unpublished March report, McGill University, School of Architecture

Sim D (2019) Soft city: building density for everyday life. Island Press, Washington, DC

Chapter 6
Orienting for the Elements

Abstract Buildings can be oriented so that their occupants will enjoy a maximum amount of sunlight. In the northern hemisphere, it will not only contribute to the reduction of heating bills but also improve the occupant's mental health by increased exposure to sunlight. In some directions, they will also benefit from a shorter noonday shadow than what would be cast if they were oriented east-west. Moreover, additional structures can be built upon the cusp of a neighboring one's shadows if the design follows the aforementioned layout. Principles considering sunlight and shadow orientation need to avoid obstructions that will prevent buildings from maximizing sunlight, a process that begins by the proper orientation of roads, studying the sun path and arranging buildings to avoid casting shadows on one another. Principles considering adaptability to existing natural conditions will be discussed in this chapter.

6.1 Considering a Place's Topography

Any kind of development involves a change in the natural form of the land. Grading a site refers to the physical adjustment to this topography and should be done with sympathy for the natural setting, particularly in cases where the site of the development is on hilly or sloped land (Holden and Liversedge 2011).

A site's original topography is a key element in the local ecosystem, holding its own distinctive flora and fauna. Thus, maintaining the topography of a site is crucial in preserving the ecosystem. To maintain and adapt to the existing topography, roads and buildings should be sited with respect to slopes and changes in elevation and must follow contour patterns. Steep slopes should also be avoided for ease of construction and visual effects (Fig. 6.1). Working with the existing terrain and natural slopes reduces construction costs and also lends to infrastructure which contains natural bumps and bends, which in turn lends to more cautious driving, increasing road safety for residents (Friedman 2012).

An effective way of arranging buildings on a sloped site is by using a terracing technique (Fig. 6.2). By arranging buildings in varying rows along the elevation, footholds—or a base for a solid foundation—can be created. By arranging buildings

A. Friedman, *Fundamentals of Sustainable Urban Design*,
https://doi.org/10.1007/978-3-030-60865-1_6

Fig. 6.1 Buildings should be sited with respect to slopes and changes in elevation and follow contour patterns. Very steep slopes should be avoided for ease of construction, cost saving and visual effects

Fig. 6.2 Buildings can be arranged on a sloped site using a terracing technique as is the case in Lisbon, Portugal

along the slope in a terraced sequence, views from each building will be unobscured. Any space left in between succeeding rows allows room for private outdoor space, decreases the obstruction of views, and prevents shadows. A sloped terrain also allows for a simple, natural, and effective drainage system, reducing cost.

Grading a site poses harmful effects on the local ecosystem: flora and fauna is uprooted and roots of trees and shrubs that run deep into the ground can easily be affected by soil compaction or water loss. It is possible to maintain the site conditions by grading in a way which incorporates the necessary structures such as retain-

Fig. 6.3 Maintaining the site conditions by constructing retaining walls, terraces and other structures for stability will reduce chance of soil erosion

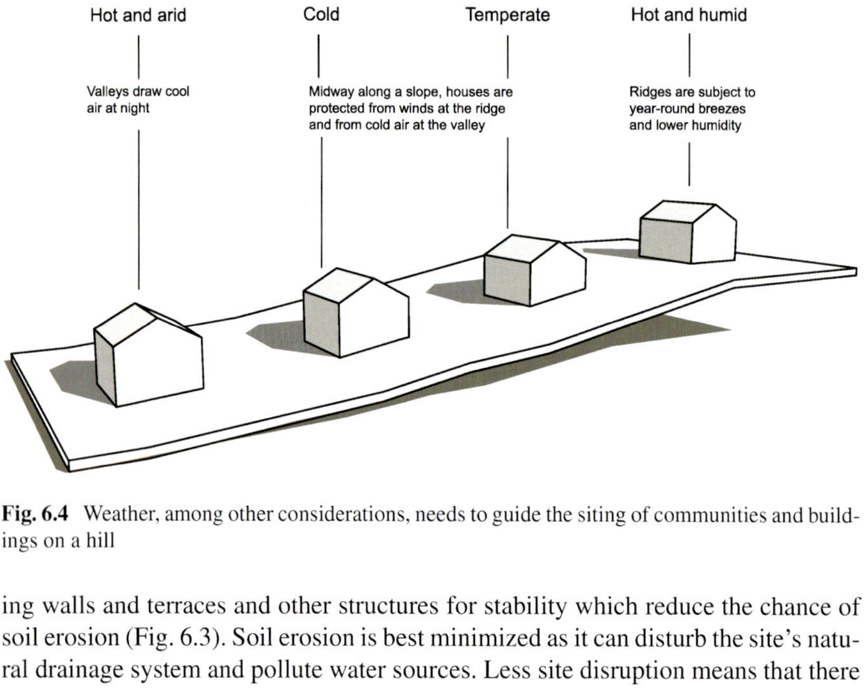

Fig. 6.4 Weather, among other considerations, needs to guide the siting of communities and buildings on a hill

ing walls and terraces and other structures for stability which reduce the chance of soil erosion (Fig. 6.3). Soil erosion is best minimized as it can disturb the site's natural drainage system and pollute water sources. Less site disruption means that there will be reduced flow of storm water to the nearby collection sites and more water will be absorbed by existing flora on the site.

It is important to consider the position of a building with respect to the slope or hill as this will largely determine the weather effects it will be subject to (Fig. 6.4). In cooler or temperate climates, it is important to shelter the buildings from cold wind. Moreover, as wind velocity increases at the ridge of a hill and cold air sinks to the bottom of a valley, it is best to locate a house halfway between the valley and the ridge to avoid both undesired extremes. On the other hand, for hot, arid climates, cold air, which settles at the bottom of the valley, is desired, and therefore buildings in such places should be located at a lower level. In addition, in warm or humid climates where more wind and air circulation are desired—and thus cross-ventilation imperative—the dwelling should be sited on a hill's ridge to take advantage of strong winds.

In general, a building should adhere to the site's existing natural conditions and fit accordingly with any existing buildings or infrastructure. Finding a suitable site by considering all environmental factors relating to elevation and topography also means considering the effects and potential gains from related weather conditions, such as wind, light, or temperature.

6.2 Orienting for Passive Solar Gain

In the same way that considering topography can greatly impact the effects of weather and the environmental ecosystem on a dwelling, considering the orientation respective to the sun is equally important. Societies have oriented their towns and homes to maximize winter sun and summer shade for thousands of years (Fig. 6.5). For example, the layouts of Ancient Mesopotamian cities were designed such that they maximized summer shade and produced a cooler microclimate within the city and its houses (Shepperson 2009). That said, with the development of efficient central cooling and heating systems in modern construction, coupled with the availability of cheap energy, society has lost interest in orientation as a key factor—losing

Fig. 6.5 Ancient Greeks and other indigenous societies oriented their towns and homes to maximize exposure to winter sun and summer shade as was the case in Linos, Greece

interest in siting dwellings for passive solar gain. With an increased awareness and sense of urgency regarding sustainable development, now is an ideal time to understand the benefits and methods of proper building orientation and implement them. When sun exposure is thoroughly considered, energy consumption can be significantly reduced, and more natural light can enter the building. As outlined by Mark DeKay, this concept is also applicable at a larger scale: if we are to shift toward renewable forms of energy over the next century, then we must begin the transition in our cities now by building physical patterns that form the infrastructure for the provision of light to buildings (2010).

The location and orientation of buildings can maximize the capture of solar energy year-round, particularly during the winter season. A dwelling should be oriented to receive the most direct sunlight during the day, with its elongated side facing south. It is imperative that the dwelling be oriented toward true south and not magnetic south, which can differ by up to 20 degrees (Fig. 6.6) (Ahmed and Sulaiman 2003). To find true south, the location's declination must first be determined. The declination is the number of degrees east or west between compass (or magnetic) north and true north. It can be found by comparing a compass reading to plans prepared by a land surveyor. Once the declination is decided on, true south can be determined. Having the elongated side of the dwelling along the east-west axis means that it can be reached by the low-angle rays of the winter sun and conversely, placing the short sides of the building to the east and west to minimize solar gains during the hot periods of summer (Numbers 1995). Furthermore, the dwelling should be sited where it lies in unobstructed sunlight, thus, placing a dwelling too close to adjacent objects or buildings should be avoided. Not only should the elongated side be exposed to the south, to fully benefit from this orientation, more windows should be placed along this axis, as the heat acquired through passive solar gain far exceeds any heat loss from the windows of that façade.

Existing natural vegetation and new growth can also be shading devices and used in a way to improve energy consumption. To minimize the solar heat gain in the summer, deciduous trees should be planted to the south of the house. The leaves will shade houses from the heat in the summer but allow some light to penetrate indoors

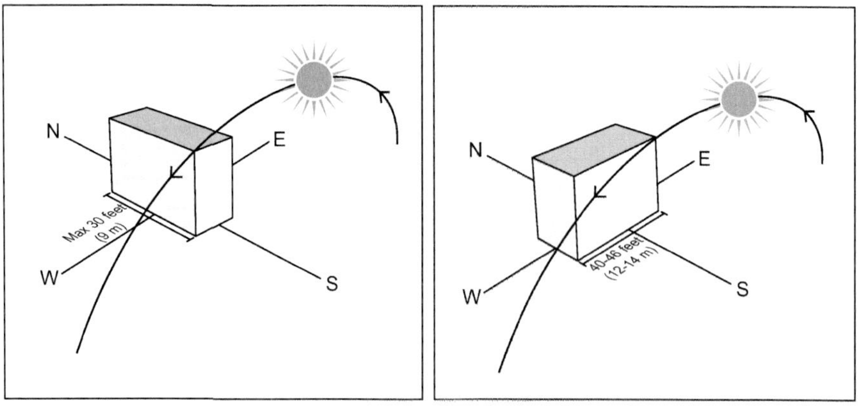

Fig. 6.6 It is highly recommended that buildings be oriented toward true south

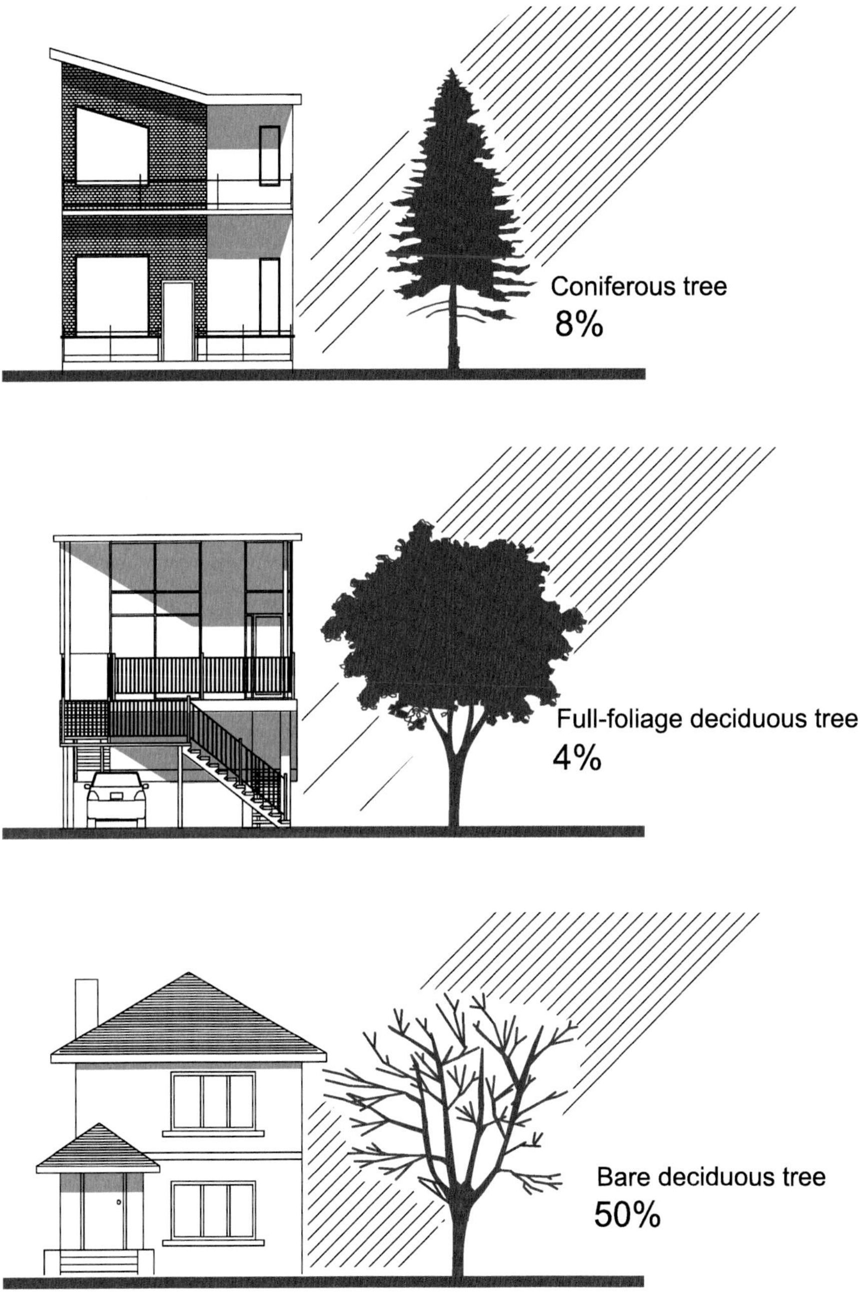

Fig. 6.7 The leaves of these trees shade houses from the heat but allow some light to penetrate. In the wintertime, more solar heat and light can filter through their bare branches

(Fig. 6.7). In the wintertime, more solar heat and light can filter through the bare branches (Friedman 2007). The east and west facades are exposed to sun with lesser intensity but should be about 10–15% fenestrated to allow for natural lighting.

Maximizing the solar gain will lower energy consumption which in turn lowers costs and reduces the amount of emitted pollution necessary to produce this energy to the environment (Mendler and Odell 2000).

6.3 Orienting for Passive Cooling

Incorporating the existing environmental elements of the site to promote natural ventilation is another important aspect of sustainable design. Trees, vegetation, and the existing landscape play a crucial role in determining wind flows. For example, near bodies of water, wind will flow from the direction of the water toward the land in the daytime but will reverse in the night. Because of inertia, wind will flow around objects and keep the same direction of flow. As a result, the vegetation in the surrounding area can greatly reduce the wind speed. During the cold months, trees act as a shield from the winter winds that generally originate from the north. Consequently, coniferous trees to the north of houses are appropriate as they maintain their foliage in the winter (Fig. 6.8). This also allows the houses to benefit from south-eastern summer breezes. Favorable positioning will allow for cross-ventilation

Fig. 6.8 Trees can act as a shield from winter winds that usually originate in the north. Planting coniferous trees to the north of buildings are appropriate as they maintain their foliage in the winter

so that mechanical cooling means will not be necessary. Considering the existing wind flow means there will be less reliance on mechanical and resource-intensive heating and cooling systems, thus making the development more sustainable.

Trees should be at a balanced distance from a building in order to give an appropriate amount of shade and also protect against strong winds. The preferred distances of woodland windbreaks from houses should be approximately ten times the tree to house height ratio. Trees planted at too great a distance from houses allow wind tunnels to sweep through the area, which are best avoided; which can also help be achieved by staggering houses and having gentle curves in the streets. Other windbreaks include architectural elements on the facades of buildings themselves such as sills and cornices.

In conclusion, the preferred building site for a new development is where the topography, sun, and wind are considered so that the benefits of the elements are maximized for energy efficiency and so that homes and buildings fit seamlessly into the landscape. When the sun and wind are considered, and trees and vegetation are planted around and in between homes, the climate can be utilized to decrease heating and cooling costs. Furthermore, properly integrating homes with their physical conditions will not only create a sustainable environment but also foster a sense of place and create a development which fits naturally into its surroundings.

References

Ahmed MM, Sulaiman M (2003) Design and proper sizing of solar energy schemes for electricity production in Malaysia. In Proceedings. National power engineering conference, Bangi, pp 268–271

DeKay M (2010) Daylighting and urban form: an urban fabric of light. J Archit Plann Res 27(1):35–56

Friedman A (2007) Sustainable residential development: planning and design for green neighbourhoods. McGraw-Hill, New York

Friedman A (2012) Fundamentals of sustainable dwellings. Island Press, Washington, DC

Holden R, Liversedge J (2011) Construction for landscape architecture. Laurence King Publishing, London

Mendler S, Odell W (2000) The HOK guidebook to sustainable design. Wiley, Hoboken

Numbers MJ (1995) Siting a house. Fine Homebuilding 93:40–45

Shepperson M (2009) Planning for the sun: urban forms as a Mesopotamian response to the sun. World Archaeol 41(3):363

Chapter 7
Space Defining Buildings

Abstract The merits of establishing mixed land uses and well-located amenities are well known and will further be elaborated in Chap. 9. The next step is how to implement and maintain these strategies. The methods can be organized into categories based on the configuration and spatial organization of the buildings themselves. This may include conversion of large-scale retail centers into transit nodes for example. This chapter discusses and illustrates how a buildings' form can define open public spaces and how such spaces can relate to one another to form a coherent urban environment that facilitates active mobility and social interaction.

Keywords Amenities · Mixed land uses · Layering of building · Sense of community · Grouping buildings · Social interaction · Public space · Communal spaces · City squares · Main streets · Outdoors areas · Urban fabrics · Complete place

7.1 Categories of Buildings' Forms

As noted above, a sustainable urban development needs to promote mixed land uses and locate amenities in an efficient, purposeful and innovative way as was the case in the design of Markthal (Market Hall) in Rotterdam, the Netherlands (Fig. 7.1). Then comes the question of arranging buildings within these mixed-use areas to define spaces. As noted by David Sim, a variety of building typologies and arrangements of in-between spaces can meet the same built density level (2019). This means that very different building sizes and layouts can meet the same density requirements. This leads to the question: what should be considered in determining building typology?

In general, each blocks' configuration offers a different collection of services and amenities: from commercial to residential units (Fig. 7.2). Bearing in mind that it is good practice to mix land uses as much as possible, the same applies to the building level. This is something which Sim elaborates throughout his work: he argues that instead of stacking, buildings should be layered (Sim 2019). Rather than one building fulfilling the same function at all vertical levels, each level should be optimized and used for its best purpose and potential (Fig. 7.3). Each level may have different

A. Friedman, *Fundamentals of Sustainable Urban Design*,
https://doi.org/10.1007/978-3-030-60865-1_7

Fig. 7.1 A sustainable urban development design needs to promote mixed land uses and locate amenities in an efficient and innovative way as was the case in the design of Markthal (Market Hall) in Rotterdam, the Netherlands

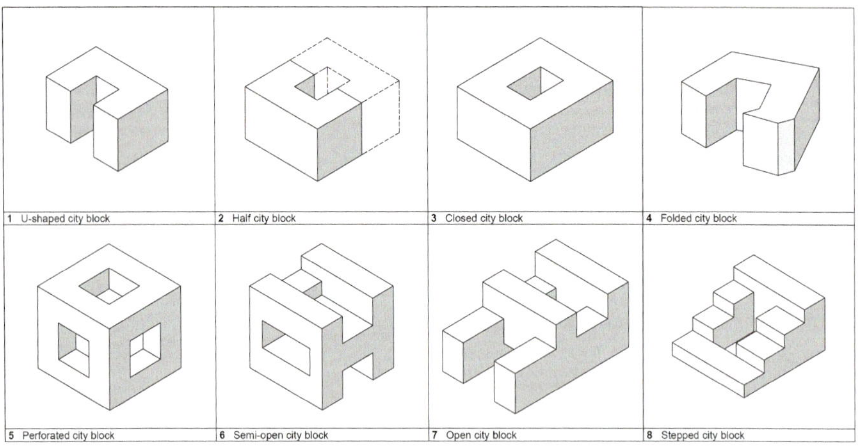

Fig. 7.2 Each city block's configuration can accommodate a different mix of amenities and residential units

advantages: the ground floor offers direct access to the street and increased visibility, on the upper floors there is potentially more natural light and privacy and on the top floor, a terrace can be of great commercial value.

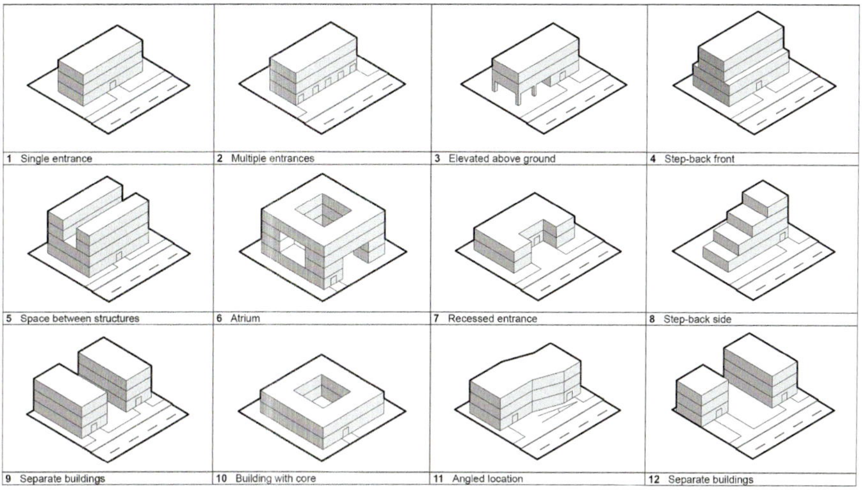

Fig. 7.3 In sustainable planning each building's level should be optimized and use to its best potential

Fig. 7.4 When more people are using the street and overseeing a movement, activity, and potential danger—and are ready to act in case of trouble—the streets become safer

The layering of building use across storeys is a characteristic seen in many sustainable neighborhoods; buildings often have shops on the ground floor and residential units above, creating multifunctional spaces. Once again, as previously outlined in this book, the tendency toward more sustainable cities is simultaneously innovation and a shift back to historical urban configurations when land uses were mixed.

Furthermore, having a space where there is both commercial and residential activity ensures that it is vibrant and full of character for a greater portion of the day. Communities with mixed-use buildings see workers arise early to start the day, children go to school, pedestrians socialize, and workers come home in the evening; a community should offer different but complementary uses throughout the day (Coupland 2013). In addition, promoting mixed-use neighborhoods not only keeps areas busy and lively, but also reduces vacant spaces. Coupland argues that these mixed-use areas can also lead to a reduced crime rate, creating safer spaces for residents (Fig. 7.4) (Coupland 2013).

To minimize travel, increase active mobility, maximize proximity to services, increase safety and foster a sense of community—the advantages of a mixed-use neighborhood—are also the reasons to have a mixed-use building. Therefore, promoting mixed-use at the building level is important and considering building form is equally important during that process.

7.2 Combining Forms

It is important to assess the different buildings—each with their own purpose—in relation to one another to create a "complete place." In complete places civic institutions such as schools, libraries, or town halls are integrated among residential and commercial buildings (Fig. 7.5). To prioritize convenience, amenities should be

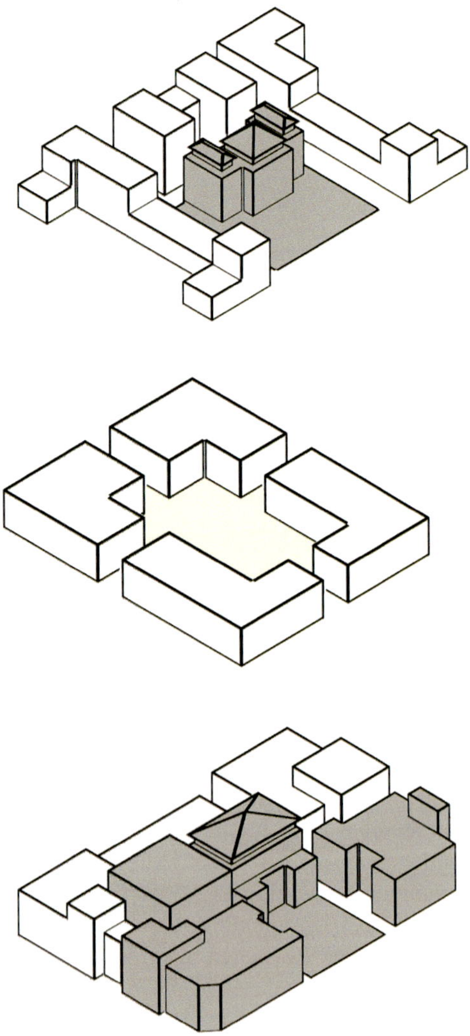

Fig. 7.5 In a "complete place" civic institutions such as schools, libraries or town halls are embedded among residential and commercial buildings

Fig. 7.6 Some buildings make room for an interior courtyards space, or a semi-enclosed space between them as is the case in Crotona, Italy (left) and Haarlem, the Netherlands (right)

grouped together to ensure efficient access to them, and in doing so, promote social interaction among the community.

When grouping buildings, enclosures—such as a large square or a small courtyard—are created. Some buildings make room for an interior courtyards space, or a semi-enclosed space between them (Fig. 7.6). Creating these types of enclosures establishes a sense of privacy in outdoor space, which is a relevant habitation form in the urban context (Sim 2019). Such a space is ideal for those who frequent the area and prefer a more secluded space. For instance, children can play outdoors near their residence in a safely enclosed area. This further contributes to the idea of creating a complete space, one within which all living and social needs are met through the creation of spaces which allow for different forms of social interaction. Additionally, assuring that these enclosed spaces are well connected by active mobility routes such as pedestrian lanes and bike paths is important in the development of urban fabrics (Fig. 7.7) (see Chap. 24 for further information).

7.3 Relation Between Buildings' Forms and Open Spaces

Planning an urban development does not end when land uses have been assigned, categorized, and building typologies assessed; an equally important stage in the planning process is addressing the space in-between. Buildings set the frame for public space: outdoor areas are vital to urban life and key in any successful urban development. The layout of buildings will determine the size of these public outdoors areas, the shelter they provide and the light they let in—setting the space for subsequent landscaping and design. Alissa North argues that by incorporating green, natural elements such as trees and vegetation in these spaces, they can reach their full potential and also become multifunctional spaces (Fig. 7.8) (2013). Most importantly, well-designed public spaces and landscapes allow for what Gehl has referred to as "life between buildings," which alludes to the spectrum of activities—recreational, social, cultural—which take place in public spaces, creating meaningful and attractive communal spaces (Gehl 2011).

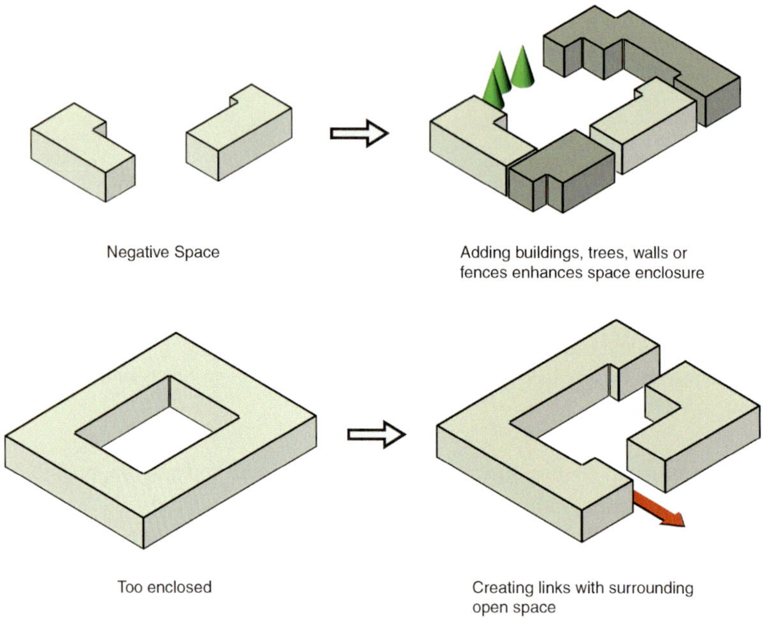

Fig. 7.7 Assuring that enclosed spaces are well connected by active mobility routes such as pedestrian lanes and bike paths is important in the overall development of urban fabrics

Fig. 7.8 By incorporating green, natural elements such as trees and vegetation in open spaces, they can reach their full potential to become multifunctional spaces

Gehl's "life between buildings" is an essential indicator of a successful sustainable development, pointing out the facilitation of social interaction. In other words, the life between buildings is at the heart of a sense of community. North argues that the quality of the public spaces—in their infrastructure, design, and natural elements—determines the investment in one's physical community or neighborhood (North 2013). When residents enjoy the communal spaces around their workplace and home, they are more inclined to use them, and build relationships with those who do. Furthermore, this leads to the creation of a sense of place in the community. It is important to outline the considerations needed to develop these public spaces so that they cater to different generational needs (see Part IV of this book for further details). Younger members of the community will require facilities such as parks, play structures, and recreational areas, whereas older members of the community will appreciate social spaces such as cafes, restaurants, or quiet benches. Bringing generations together is important; some spaces such as central city squares, main streets, or markets attract people of all ages to the same space; fostering a mutually beneficial support system in the community. In addition, it is also noted that accessibility must be ensured to cater to the needs of all community members, be it physically through ramps and elevators or economically by removing any financial barriers so that public spaces can be enjoyed by anyone regardless of their age, income, or physical ability.

Once again in discussing the creation of mixed-use areas at the land use and building level, the idea of incorporating supporting relations of sustainability is at the core. Diversifying building use and making use of outdoor spaces for various activities and social interaction—in both a more open and private way—all contributes to creating a space which is more environmentally, economically, and socially sustainable. Moreover, in creating these clusters of economic and social interaction, residents feel a sense of community and to quote Sim, a sense of "living locally in an urbanizing world" (Sim 2019).

References

Coupland A (2013) Reclaiming the city: mixed use development. Taylor and Francis, Hoboken

Gehl J (2011) Life between buildings. Island Press, Washington, DC

North A (2013) Operative landscapes: building communities through public space. Birkäuser, Basel

Sim D (2019) Soft city: building density for everyday life. Island Press, Washington, DC

Chapter 8
Sense of Place, Human Scale, and Vistas

Abstract The experience of life in a city is influenced by the space itself. Sustainable urban design must encourage participation in city life through design at the human scale, sense of place, and vistas. Human scale refers to the way we proportion our surroundings and relate our own height to the places in which we find ourselves. Comfortable scale is the outcome of well-proportioned street width, appropriate setback of buildings from the road, and the height of buildings, trees, streetlamps, and other design features. Appropriate scale enhances our sense of place and comfort. Disproportionate scale, on the other hand, can make us feel out of place and disconnected. Similarly, having long views, known as vistas, will enhance walkability. The shared benefits of human scale and vistas contribute to a city's sense of place. This chapter discusses the importance and implementation of these aspects in sustainable urban design.

Keywords Human scale · Sense of place · Vistas · Well-proportioned street width · Appropriate scale · Long views · Car-centric design · Identity · Imageability · Active modes of transportation · Urban life · Cycling culture · Street furniture · Visual corridors

8.1 Sense of Place

As explored in Chap. 1, from the twentieth century onward, planners were committed to a model of city life that prioritizes the automobile. Low-density, car-centric design has transcended suburbia and made its way into the North American urban landscape. This form of design threatens the spontaneous, engaging experience of the city. It eliminates nuance in the built environment; instead, features are designed to capture the fleeting attention of passing drivers (Kay 1997). There must be a paradigm shift in planning that re-prioritizes the pedestrian that begins by establishing a sense of place.

Places give the people who inhabit them a sense of identity; they can be engaging, inspiring, and nurturing. Urban form lends itself to distinct experiences that heighten the unique sense of place from city to city. For example, wide boulevards

A. Friedman, *Fundamentals of Sustainable Urban Design*,
https://doi.org/10.1007/978-3-030-60865-1_8

Fig. 8.1 Barcelona's wide boulevards, originally designed for military purposes, are one of the elements that lent the city its identity

in Paris and Barcelona, originally designed for military purposes, have resulted in an iconic "boulevard culture" that generates social life along the city's sidewalks (Fig. 8.1) (Gehl 2010). More generally, "sense of place" refers to the ways that people interact with and interpret their surroundings, based on how its social, environmental, cultural, or historical identity is expressed through the built environment (Adams et al. 2016).

Fig. 8.2 Urban features such as outdoor dining contributes to the local's imaginability to forester a unique sense of place

Sense of place is linked to Kevin Lynch's (1960) concept of "imageability," an aspect of the built environment that evokes a memorable image (Lynch 1960). When factors such as distinctive or historic buildings, outdoor dining spaces, and other landmarks contribute to imageability, a space is likely to have a pleasant sense (Fig. 8.2) (Ewing and Handy 2009).

Cities with a pleasurable sense of place must be designed at a comfortable scale to promote active modes of transportation and opportunities for spontaneous social interaction. However, the human dimension of urban life has historically been overlooked in favor of planning ideologies that support vehicular traffic and emphasize individual buildings rather than common spaces. This pattern has degraded the traditional functions of cities and eroded the sense of place that improves city dwellers' quality of life (Gehl 2010).

Jan Gehl makes a compelling case for the importance of a sense of place in *Cities for People* (2010). Gehl argues that prioritizing a sense of place is essential when planning a city that encourages full participation in urban life. In a city with a strong sense of place, people will walk, enjoy public spaces, and interact with features of the built environment such as parks, bicycle paths, and active ground floors. A sense of place creates a framework for urban life that connects people to the city and promotes the sensations of comfort and community that define human scale. Sense of place is created and reinforced through a dense city structure, with amenities located on the ground floors of mixed-use buildings (Fig. 8.3). Beautiful public spaces, distinctive architecture, and deliberate design encourage people to walk, bike, and participate in

Fig. 8.3 Mixed-use buildings with "active" ground floors and galleries reinforced a sense of place

the social and economic life of the city. Specific features such as cafés or promenades contribute to a city's unique atmosphere: for example, in many European cities, bicycling infrastructure results in a sense of place centered around a cycling culture (Gehl 2010).

For these design features to benefit city residents, they must be accessible within a short walking distance of one another and appropriately scaled so they can be observed and appreciated in detail.

8.2 Human Scale and Sustainable Planning

Human scale refers to the way we proportion our surroundings and relate our own height to the places in which we find ourselves (Friedman 2018). Cities that are planned with the human scale in mind are in direct opposition to the wide roads, oversized billboards, and tall buildings that characterize many modern cities, which were built to be observed out the window of a moving vehicle (Gehl 2017). A city built at the human scale should feature a well-proportioned street width, with a height-to-width ratio of 1:1, where pedestrians are prioritized over vehicular traffic (Fig. 8.4). Houses will be set back an appropriate distance from the road, and they are built so that their front doors, not their garages, are visible to pedestrians. Buildings, trees, and streetlamps should be proportionate height and are in place to grant shade, frame vistas, and increase feelings of security. Human scale enhances people's comfort and perceived safety in a neighborhood, but in the decades following World War II, North American cities increasingly expanded roads that were constructed with little regard to this important design feature (Fig. 8.5). Instead, planners have widened streets to accommodate vehicular traffic, while eliminating alleys ways and pedestrian paths. Urban designers have differing opinions on the specific dimensions that contribute to human scale, but generally, building heights should be capped below six stories, or other architectural elements are required to mediate the effects of tall building heights. Building and street width is also taken into consideration, and trees can be used to moderate disproportionate width. In addition to these elements, design features such as paving patterns, street furniture, and building ornamentation can improve scale (Ewing and Handy 2009).

Correcting the scale of post-war planning practices by improving city spaces at the human scale will lead to new patterns of use and better quality of life. Human

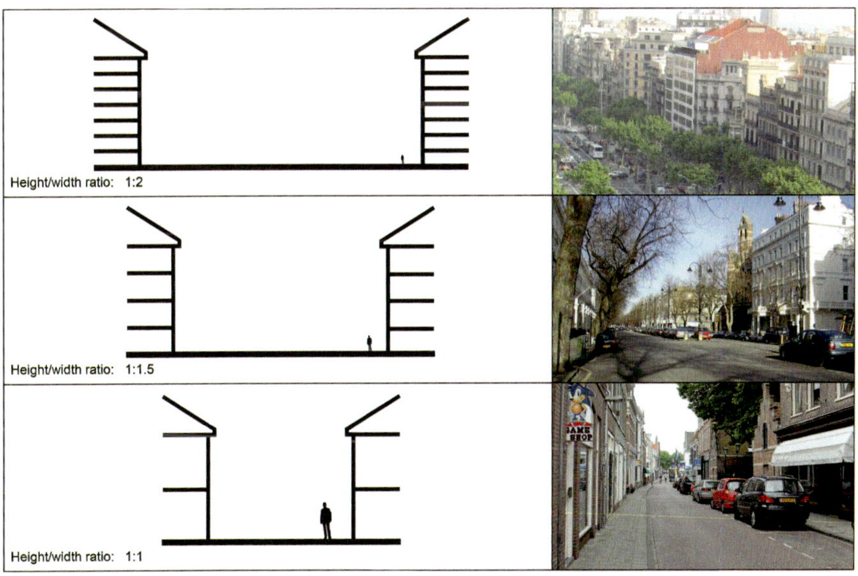

Fig. 8.4 A recommended human scale for a city features a well-proportioned street width, with a height-to-width ratio of 1:1

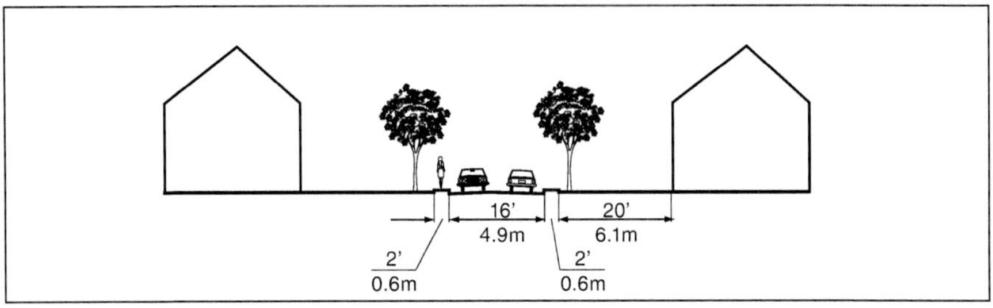

Hampstead Garden Suburb, 1905

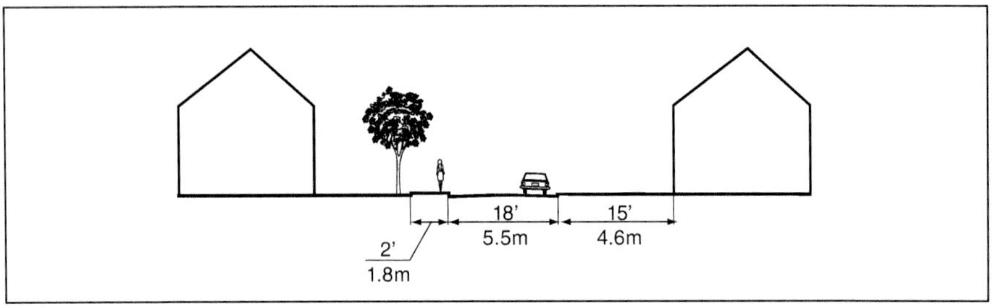

Radburn, New Jersey, 1927

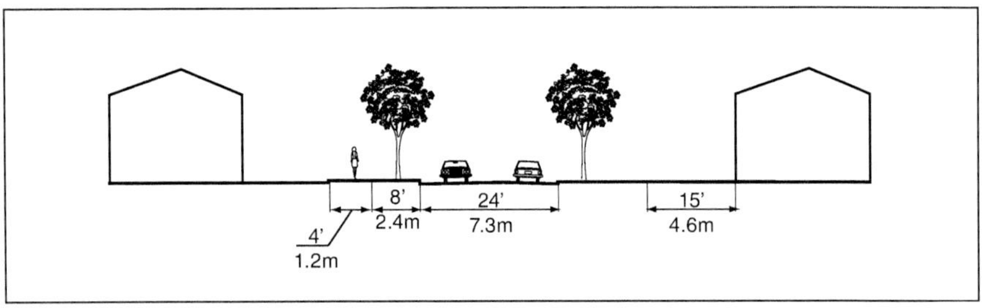

FHA Standards, 1936

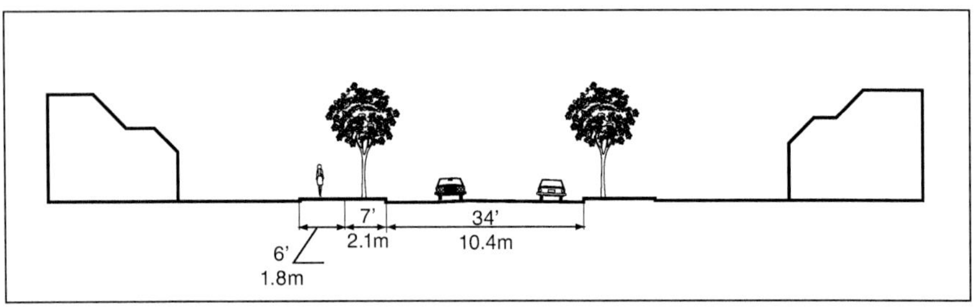

ITE Standards, 1965

Fig. 8.5 In the decades following World War II, North American cities increasingly expanded roads' widths with little regard to human scale

scale is an important perceptual quality that contributes to walkability (Ewing and Handy 2009). If cities are designed at a scale that encourages its residents to walk and take advantage of public spaces, people will engage with their surroundings in a way that encourages the formation of a more sustainable urban environment (Gehl 2010).

Human scale design contributes to sustainability in an economic, environmental, and social sense. Road maintenance, resurfacing, cleaning, and snow removal now account for 25% of a home's cost through municipal property taxes. Scaling roads appropriately will reduce these costs and make housing for example more affordable. Moreover, promoting mixed-use buildings with ground-level commerce will allow small businesses to be successful and accepted in a neighborhood, keeping services, and money local (Friedman 2018). Designing cities at a scale that encourages walking promotes environmental sustainability by reducing emissions associated with vehicular transit (Gehl 2010), while improving life expectancy and overall health. Walkability also encourages optional activities like walking down a scenic route, visiting a look-out point, or people-watching on a bench. These activities make cities pleasant and exciting places to live and can be promoted through deliberate, human-centered urban design.

Historically, zoning ordinances favor single-family homes, and more generally subordinate health and safety concerns in support of raising property values. In some municipalities, bylaws state that homes must be set a minimum of 9 m (30 ft.) from the road, eroding human scale. Therefore, efforts to make neighborhoods more sustainable require a reevaluation of municipal zoning standards, as illustrated in Fig. 8.6. Outdated zoning practices are varied in their specific requirements, but they all affect the planner's ability to design at the human scale. When cities are constructed at a disproportional scale, people feel disconnected and out of place. Countering this through planning at the human scale inherently promotes neighborhood sustainability by fostering a sense of place and an attachment to one's community.

Fig. 8.6 A proper human scale is achieved when buildings are set closer to the sidewalks as is the case in this development near Montreal, Quebec, Canada

8.3 Maintaining Vistas and Their Importance

Vistas are visual corridors that allow for a long view of an attraction at the end of a walkable path. They can be used to draw attention to a specific feature of the natural or built environment and are an important tool in improving a city's walkability (Fig. 8.7). The details of a path that features a vista—such as the number of entry points, the amount of vehicular or pedestrian traffic, and the topology—contribute to an attractive and distinctive scene. Emerging from a narrow corridor into a vista has a striking visual impact that certainly affects the pedestrian experience.

Vistas encourage people to walk for both recreational and utilitarian purposes, which improves neighborhood sustainability. They also mediate the effect of tall buildings and wide streets (Ewing 2013). When a vista-containing path is framed by mature trees, the canopy creates a feeling of unity and closeness that results in comfortable scale. Vistas allow for an unusual level of control on the part of the planner, whose preservation of a vista allows for a distinctive and memorable pedestrian experience (Friedman 2015). Long views also contribute to a sense of enclosure. The illusion of fixed boundaries in a city space makes it seem special, therefore contributing to a sense of place and improving walkability. (Ewing and Handy 2009).

Fig. 8.7 Vistas are visual corridors that allow for a long view of an attraction at the end of a walkable path

In *Cities for People*, Jan Gehl defines the "social field of vision" at approximately 100 m (328 ft.). This distance is the maximum length that enables pedestrians to experience the vista and enjoy an overview of the space in front of them. The long view should be balanced with small-scale elements meant to be enjoyed in detail by a pedestrian walking 4–5 km per hour (two to three miles per hour) (Gehl 2010).

Vistas must be delicately preserved from conflicts that arise from new construction. The particular path gradients that create vistas are highly specific and thus these views can be easily blocked by vehicular pathways or other urban developments. The presence of a vista can attract pedestrians to a certain area and should therefore be planned for and accommodated during later planning stages.

Urban design that includes an appropriate and comfortable sense of proportion leads to more sustainable communities. Designing cities at the human scale reinforces the sense of place that make cities distinctive, vibrant, and pleasant places to live. Although features like walkability, scenic paths, well-proportioned roads, trees, streetlamps, and vistas bring a necessary human dimension to urban design, planners must navigate outdated zoning bylaws and decades of automobile-centric practices to implement a human scale and create more sustainable cities.

References

Adams J, Greenwood D, Thomashow M, Russ A (2016) Sense of place. https://www.thenatureofcities.com/2016/05/26/sense-of-place/. Accessed 22 Jan 2020

Ewing R (2013) Eight qualities of pedestrian- and transit-oriented design. Urban Land Institute and American Planning Association, Washington

Ewing R, Handy S (2009) Measuring the unmeasurable: urban design qualities related to walkability. J Urban Des 14(1):65–84

Friedman A (2015) Planning sustainable neighbourhoods. Springer, Cham

Friedman A (2018) Neighbourhood: designing a liveable community. Véhicule Press, Montréal

Gehl J (2010) Cities for people. Island Press, Washington, DC

Gehl J (2017) In the last 50 years architects have forgotten what a good human scale is. https://www.archdaily.com/877602/jan-gehl-in-the-last-50-years-architects-have-forgotten-what-a-good-human-scale-is. Accessed 22 Jan 2020

Kay J (1997) Asphalt nation. University of California Press, Berkeley

Lynch K (1960) The image of the city. MIT Press, Cambridge

Chapter 9
Mixing Land Uses and Dwelling Types

Abstract The segregation of land use into residential, commercial, and industrial zones was introduced in the early twentieth century. Known as Euclidean zoning, it indirectly gave way to the rise of suburbia, encouraged automobile dependency, limited walkability, and eroded social diversity. Neighborhoods with mixed land uses and a variety of dwelling types help curb urban sprawl. This is to say nothing of the fact that they offer richer built environments and social opportunities to their residents. This chapter illustrates the merits and modes of mixing a variety of housing types and amenities in cities to create better urban environments.

Keywords Euclidean zoning · Urban sprawl · Automobile dependency · Sustainable cities · Diversity of land uses · Housing types · Mixed communities · Social cohesion · Affordable housing · Sustainable urban life · Transit nodes · Retail hubs · Transit-Oriented Developments (TOD) · Pedestrian Pockets

9.1 Mixed-Use Planning and Neighborhood Sustainability

In 1926, the Supreme Court in the USA found zoning ordinances to be constitutional as long as the ordinances improve public welfare (Oyez n.d.). This landmark case, Village of Euclid v. Ambler Realty, ushered in an era of segregation by land use. Unlike streets in traditional cities, this practice, which is known as Euclidean zoning, defines the contemporary North American landscape (Fig. 9.1). Zoning is not inherently bad; separating industrial and residential land is a beneficial practice to mitigate the dangers of pollution (Watsky 2018). However, Euclidean zoning in some cases has contributed to the creation of communities that are racially and socioeconomically segregated, and car dependent (Hall 2007). Urban designers must adopt alternative practices that recognize the importance of mixed dwelling types in the creation of sustainable cities.

Euclidean zoning has institutionalized urban sprawl, a phenomenon characterized by low-density development and limited walkability (Hall 2007). Cities with a mix of dwelling types and land uses also help curb urban sprawl by increasing density, while improving residents' quality of life through walkability for example.

A. Friedman, *Fundamentals of Sustainable Urban Design*,
https://doi.org/10.1007/978-3-030-60865-1_9

Fig. 9.1 In the heart of traditional cities people still reside above commercial establishments as is the case in Evora, Portugal

This can be achieved through the construction of multiunit dwellings such as apartments, duplexes, and buildings with businesses on the ground floor (Fig. 9.2). Mixing also includes increasing the diversity of land uses so that compatible uses will be brought together. For instance, locating different dwelling types near community institutions so that residents can access these institutions more easily. Finally, mixing land uses requires the integration of previously segregated zones, which often involves overcoming regulatory barriers (Grant 2002). Mixing a variety of housing types and amenities encourages built form and social diversity. A neighborhood's density, maximized through mixed land use, affects its economy, its social fabric, and as a result increases its potential to become sustainable. The negative social outcomes associated with Euclidean planning can be addressed through varying land use, given that this form of development opens communities to a wider range of people and businesses.

As urban areas are becoming denser, there is an increasing number of seniors, young people, and families with children. Mixed communities offer these groups stimulating, active, convenient, and social lives. Living in a diverse neighborhood with civic, community, and business institutions nearby is beneficial to all residents. It is of particular importance to those who work from home, which represents a growing proportion of the workforce in the age of digital communications.

Without mixed-use planning, social segregation and its associated ills are exacerbated: if there are no businesses in a wealthy residential neighborhood, outsiders have no reason to visit. This erodes social cohesion between different socioeconomic and ethnic groups. Municipalities legitimize segregation using numerous

Fig. 9.2 The vertical layering spaces in Haarlem's buildings in the Netherlands have stores on the ground level, offices above them topped by dwelling units

zoning tactics. Many towns exclude low-income residents by mandating that homes must be built on large lots and that homes be constructed in a particular style. Others are zoned exclusively for single-family residences and prohibit the construction of apartments and duplexes (Hall 2007). Mixing dwelling types and amenities, on the other hand, fosters inclusive and diverse communities as was demonstrated in Västra Hamman near Malmö, Sweden (Fig. 9.3) (Feehan and Feit 2006). Planning for mixed land use can be used to provide affordable housing. When more dwellings are built on the same plot, the cost of land decreases. Not only does increasing density reduce the price of the home itself, it divides the costs associated with homeownership (such as maintaining water lines, roads, and trees) across more units, making the neighborhood more accessible (Friedman 2018).

Mixed-use planning—and an increase in density—has the potential to create more open spaces (Hall 2007). Varying land uses allows for community gardens, parks, and other green spaces to be integrated alongside residential and commercial developments, allowing even distribution (Fig. 9.4). Green spaces are of great ecological, social, and economic importance. They disrupt the urban heat island effect, encourage an active lifestyle, preserve biodiversity, and improve food security. Diverse communities benefit from shared outdoor space through an increased sense of cohesion (Friedman 2017). Integrating public open spaces amidst different dwelling types will encourage people to interact with their neighbors and engage with their surroundings in a more meaningful way.

Fig. 9.3 Having a mix of dwelling types and amenities, fosters inclusive and diverse communities as was demonstrated in Västra Hamman near Malmö, Sweden

Fig. 9.4 Varying land uses, and dwellings' types allows for parks and other public green spaces to be distributed evenly

Mixing land uses has been instrumental in the revitalization of industrial corridors in several cities (Grant 2002). Businesses benefit from proximity to people; without users nearby, commerce will be less economically viable. An interesting mix of residential, office, commercial, and entertainment spaces draws visitors from outside the neighborhood as well (Grant 2002). Ideally, this arrangement will proliferate the growth of small businesses and keep residents' money local, allowing the neighborhood to achieve economic self-sufficiency.

The promise of economic investment and neighborhood reanimation must be balanced with an equitable mix of dwelling types: often, in embracing mixed land uses in a former industrial area, municipal officials are quick to directly or indirectly displace former residents (Grant 2002). To achieve the equitable, inclusive, and sustainable urban life that is associated with mixed land uses, planners must incorporate diverse housing types that are accessible to a range of socioeconomic groups.

9.2 Planning for Mixed Units, Heights, and Densities

Planning methods for mixed uses can be organized into categories based on neighborhood configuration, spatial organization, transit nodes, and retail hubs as illustrated in Fig. 9.5. When planning for mixed dwelling types, designers must consider all aspects of a neighborhood: the demographic criteria, the built environment, and the aesthetics of the surrounding areas. A well-designed mixed-use neighborhood will encourage its residents to walk, take transit, make use of community organizations, and participate in the local economy.

"Neighborhood configuration" refers to the design and connectivity of different landscape elements. Considering neighborhood configuration can optimize the benefits of mixed-use planning (Hersperger 2006). A planning strategy can be the creation of transition nodes, an arrangement in which smaller buildings encircle large commercial establishments to reduce traffic congestion and create a more aesthetically pleasing urban environment (Fig. 9.6). Another configuration tactic is the introduction of light industry to residential areas, when the technology involved is advanced enough to assuage concerns about pollution.

The spatial organization of a city's district contributes to its walkability; ideally, all services are located within walking distance from homes. To achieve this level of accessibility, commercial space should be placed at street level, with residences seated above stores. Urban space can be organized either horizontally or vertically (Fig. 9.7). Vertical organization uses less land than horizontal organization; in this case, walkability and density is achieved through the presence of several high-rise buildings. The high-rises act as landmarks and include segregated spaces for commercial, residential, and other uses. Research has found that the spatial organization of mixed land use cities can be instrumental in the city's ability to achieve its policy objectives (Bertaud 2004).

Ideally, the spatial organization of a neighborhood will facilitate walkability. Of course, in larger neighborhoods, this may not be feasible. In that case, centers of

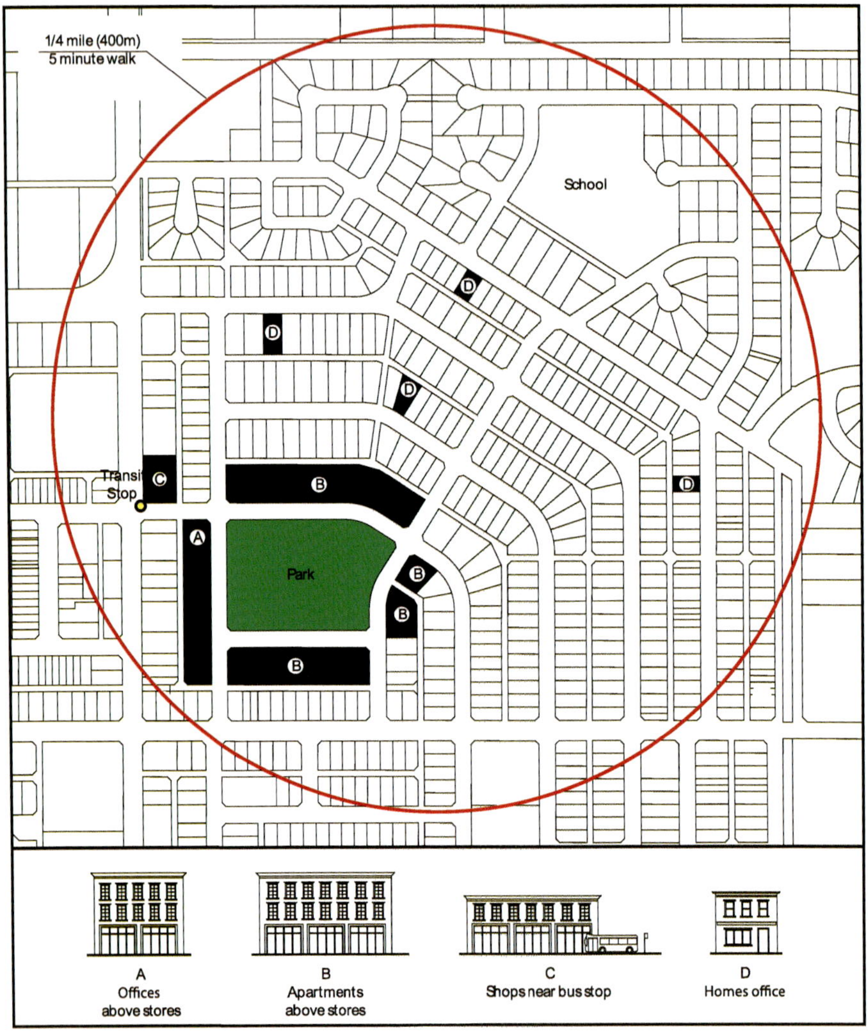

Fig. 9.5 Planning methods for mixed land use can be organized into categories based on neighborhood configuration, spatial organization, transit nodes, and retail hubs

activity should be connected by transit stops. Transit-Oriented Developments (TOD) promotes the introduction of mixed-use areas around transit hubs, each with a unique feature to draw people from other regions. Known as "Pedestrian Pockets," a space of no more than 0.4 km (0.25 mile) can be placed near transit stops, allowing easy and walkable access to amenities as illustrated in Fig. 9.8. Pedestrian Pockets can also be designed as retail corridors to attract visitors from other neighborhoods. A main shopping street with an interesting mix of shops can be a pleasant place to shop and has the added benefit of increasing foot traffic to local businesses.

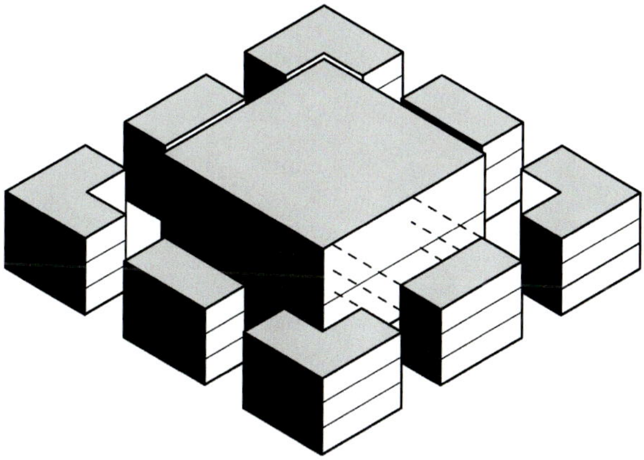

Fig. 9.6 A planning strategy is the creation of transition nodes, an arrangement in which smaller buildings encircle large commercial establishments

Fig. 9.7 Mixed-use urban space can be organized either low rise horizontally as is the case in Den Burg, the Netherlands (left) or vertically like in Brussels, Belgium (right)

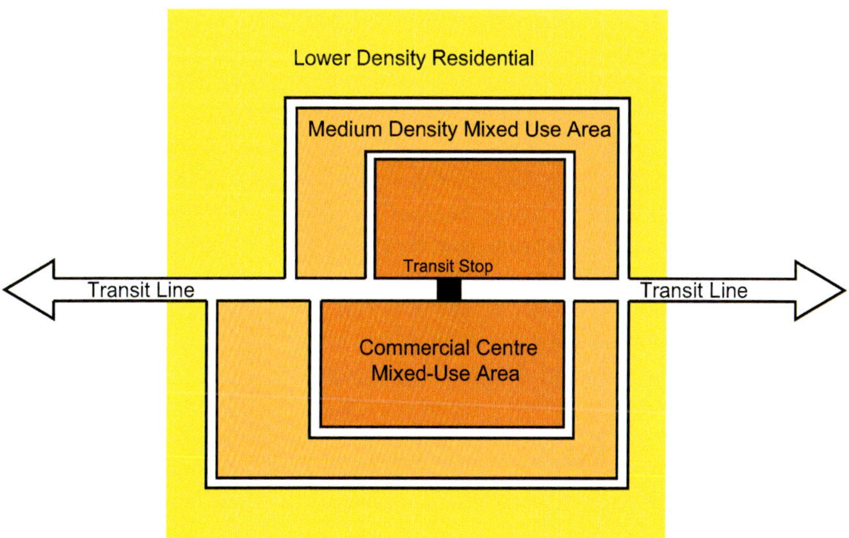

Fig. 9.8 Pedestrian Pockets can be placed near transit stops, within easy walkable access of amenities

Fig. 9.9 The Benny Farm community in Montreal, Quebec, Canada features a variety of dwelling typologies to make it affordable and foster a rich mixed population

When developing new communities with mixed land use, designers must consider the dwelling typologies and architectural motifs of the neighborhood, as well as the demographic criteria of the planned community. Benny Farms is a community in the Notre Dame des Grace neighborhood in Montréal, Canada that features a variety of dwelling typologies (Fig. 9.9). The surrounding neighborhood is composed of single-family homes and duplexes, mostly under three storeys high. Balconies are a key architectural motif. Benny Farms itself was designed to include a wide range of dwelling types to accommodate a diverse demographic. Units were tailored to specific budgets and were designed so adjacent units could be retroactively fit together. Different demographics are grouped together: veterans and seniors share a quieter zone, young families are in another zone, giving children space to play together, and all share easy access to mutually beneficial services.

Benny Farms is just one example of a well-designed community. The ideal mixed-use neighborhood varies from city to city; however, there are several key features that optimize the urban environment. First, residences should be located near a commercial center. Allowing a mix of dwelling types, from single-family homes to apartments, can sustain several commercial amenities and make the neighborhood a more convenient and enjoyable place to live. The neighborhood should contain civic institutions such as government offices, day cares, and post offices that cater to the needs of a diverse population. Similarly, mixed-use communities should include convenience centers, which provide necessary services and increase social cohesion. The purpose of these clustered amenities is to support the needs of a socioeconomically and racially diverse group of residents, while encouraging an active lifestyle. In designing neighborhoods, planners must push back against outdated Euclidian zoning bylaws to establish more sustainable mixed-use communities.

References

Bertaud A (2004) The spatial organization of cities: deliberate outcome or unforeseen consequence? https://escholarship.org/uc/item/5vb4w9wb. Accessed 1 Feb 2020

Feehan D, Feit M (2006) Making business districts work. Routledge, New York

Friedman A (2017) Designing sustainable communities. Bloomsbury, New York

Friedman A (2018) Neighbourhood: designing a liveable community. Véhicule Press, Montreal

Grant J (2002) Mixed use in theory and practice: Canadian experience with implementing a planning principle. J Am Plan Assoc 68(1):71–84

Hall E (2007) Divide and sprawl, decline and fall: a comparative critique of Euclidean zoning. Univ Pittsb Law Rev 68(4):915–952

Hersperger A (2006) Spatial adjacencies and interactions: Neighbourhood mosaics for landscape ecological planning. Landsc Urban Plan 77(3):227–239

Oyez (n.d.) Village of Euclid v. Ambler Realty Company. https://www.oyez.org/cases/1900-1940/272us365. Accessed 1 Feb 2020

Watsky R (2018) The problems with Euclidean zoning. http://sites.bu.edu/dome/2018/07/19/the-problems-with-euclidean-zoning/. Accessed 1 Feb 2020

Chapter 10
Infill Projects and Strategies for Their Integration

Abstract Vacant city land offers opportunities and challenges to urban designers. Proximity and easy linking with existing roads and utilities, and contribution to the rehabilitation of rundown neighborhood are some valuable advantages. Encountering contaminated soil, odd-sized lots, and facing Not in My Backyard (NIMBY) sentiments are some of the drawbacks. Yet, the need to curb urban sprawl, adopt smart growth strategies, and lower housing costs makes infill housing worth pursuing. This chapter elaborates on issues related to urban design of infill projects and dwells on strategies for their physical integration.

Keywords Not in My Backyard · Urban sprawl · Smart growth · Infill housing · Physical integration · Low-density · Automobile-dependent growth · Active transport · Social cohesion · Live-work housing · Brownfield sites · Affordable housing · Passive solar gain

10.1 The Importance and Challenges of Infill Projects

Infill projects be they residential, commercial, or institutional are built on vacant land within an already developed area (Fig. 10.1). These projects serve the valuable purposes of avoiding urban sprawl and improving the areas around existing development. They also reduce infrastructure costs, since public utilities are already in place near the development. However, there are many challenges that threaten infill projects' success.

Urban sprawl, a form of low-density, automobile-dependent growth exacerbates social, economic, and environmental ills (Squires 2002). The ills of urban sprawl have long been understood by planners, however, the problem persists. Infill development is a promising alternative that will minimize sprawl by taking advantage of land in already developed urban and suburban areas. Selecting these sites for development will bring the benefits of mixed land uses and will decrease the negative externalities associated with suburbanization.

Infill development has the added benefit of improving the area around which development occurs. This was the case in the town of Peace River, in Alberta, Canada.

A. Friedman, *Fundamentals of Sustainable Urban Design*,
https://doi.org/10.1007/978-3-030-60865-1_10

Fig. 10.1 Infill housing projects, be them residential, commercial, or institutional are built on vacant land within an already developed area like this one in Edam, the Netherlands

The town faced a series of challenges that eventually led to its decline, which called for a strategy of regeneration. First, a new commercial center was built on the edge of town, which forced several businesses in the core to close or move to the new commercial area. The core of Peace River suffered as a whole from the development of large commercial stores. Another area of concern for the town was their dependence on few key industries. Their economic base needed to be broadened which led to the exploration of a tourist industry that could take advantage of the area's natural beauty. There was also a desire to improve Peace River's sense of place and to create natural meeting places for both residents and visitors. Additional barriers in the city include stagnant population growth, lack of human scale, outdated businesses giving the place a poor image, and lack of access to the river (Fig. 10.2).

The same processes that led to urban sprawl resulted in decreased investment in inner-city areas, a disparity that falls along racial and socioeconomic lines. The departure of industry from the inner city, along with demographic changes beginning with "white flight" in the 1950s, have resulted in societal and economic changes that make infill housing projects in the urban core highly relevant. Attracting new residents to a neighborhood will broaden its tax base and attract new businesses. Providing a mix of housing types will make the neighborhood accessible to residents with a wider range of incomes, who will choose to live in the same neighborhood. The diversity will create a vibrant and welcoming street scene in which residents are excited to participate.

Fig. 10.2 Infill development has the added benefit of improving the area around which development occurs as was the case in the town of Peace River, Alberta, Canada

When permitted by zoning laws, infill housing provides the opportunity for mixed-use projects. Mixing land uses increases the diversity and density in the built environment, making the neighborhood more sustainable by encouraging active transport and social cohesion. Diversity and density also promote a sense of safety, further improving urban communities (Carswell 2012).

Infill projects also address a need in the real estate market. Often, the market for infill projects is generated by increased employment opportunities in a central business district as was the case in the town of Fort Saskatchewan, in Alberta, Canada. As illustrated in Fig. 10.3, the chosen strategy paid close attention to the back alleys in the town's center, specifically by converting them into local pedestrian streets through which people could access new innovative live-work housing projects on infill lots. Those housing were meant to attract young people and start-up companies.

Access to transit, work, and cultural experiences encourage certain demographic groups to seek housing close to the city center (Farris 2001). For instance, single people, single-parent families, and the elderly are more interested in smaller, lower cost dwellings built in higher density settings, especially when their residences are close to amenities and transit.

An added benefit of infill projects is the ability to spend less on public utilities and servicing since they are already in place. Additionally, it is more economically viable to add new public services and amenities following an infill project, because the residential intensification will necessarily bring a larger tax base to the area. However, this same process can increase strain on infrastructure like roads, sewers, and public transportation (Carswell 2012).

Unfortunately, there are many barriers to successful infill development. Many inner-city neighborhoods are not optimal candidates for infill projects due to deteriorated infrastructure, historical disinvestment, and a lack of supporting facilities and services (Farris 2001). Local opposition is common for infill projects, especially when there is affordable housing included. When construction begins, pollution and noise might affect neighboring homes and pose public health concerns. Infill development often leads to NIMBYism, a sentiment where people may support the idea behind a development but reject its presence in their neighborhood (Friedman 2005).

There are significant environmental challenges associated with brownfield sites, a form of infill housing wherein developments are built on previously used industrial land (Farris 2001). The first issue with brownfield projects is site clean-up. Often, little or no information about the type and extent of contamination on a site is known without an environmental site assessment (ESA). An ESA includes soil and water testing and lays out a plan for remediation consistent with local and national standards. Unfortunately, the remediation process may be costly, and even after it is completed, community groups may express concerns about living on brownfield sites (Wells 2007).

The size, shape, and location of the site and the architectural style of the surrounding buildings might make infill housing projects challenging to design and costly to build. Sites may be just one part of a larger plot of land, and as a result they might be oddly shaped. The lot's shape can pose logistical challenges for designers

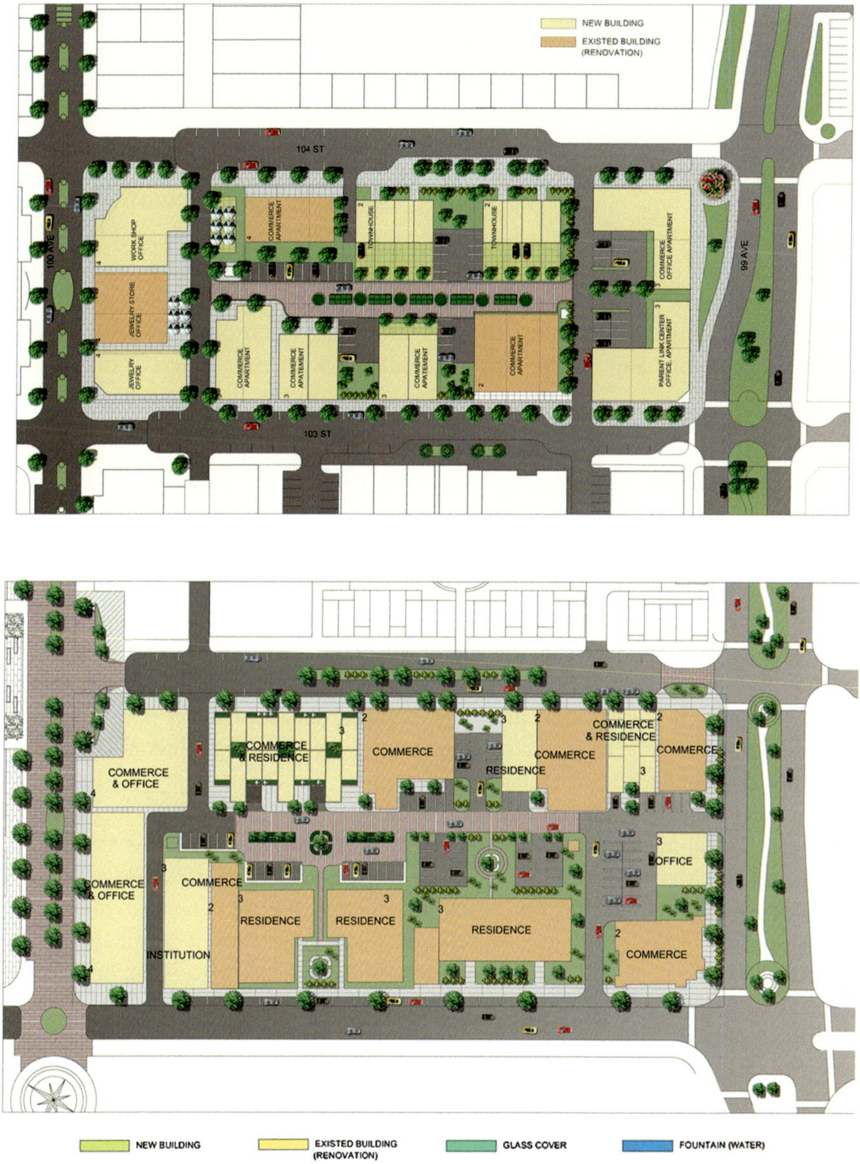

Fig. 10.3 In the town of Fort Saskatchewan, Alberta, Canada, a chosen strategy paid close attention to the back alleys by converting them into local pedestrian streets through which people could access new innovative live-work housing

and builders. Similarly, planners may need to update infrastructure to accommodate the increased population density.

Legal and political challenges abound with infill housing. If land is privately held, the developer may run into problems with occupants who do not want to sell or who ask for exorbitant prices. This problem is generally dealt with through eminent domain, wherein the local government exercises their power to take control of privately held land for public use. This solution takes time and money, and represents a significant opportunity cost for developers, who may revert to the simpler process of building on greenfield suburban land (Farris 2001). Zoning bylaws can further jeopardize the feasibility of infill housing projects. Regulations regarding lot size and dwelling type hinder infill projects (McConnell and Wiley 2010).

Finally, it is essential to consider the relationship between displacement, gentrification, and infill development. As is the case with most neighborhood regeneration efforts, developers run the risk of eroding the place's amenities and community. Planners must make income diversity a central goal, to ensure vitality, economic health, social equity, and sustainability in infill development projects (Kim 2016).

10.2 Integration Strategies and Methods for Infill Projects

The optimal infill site is located in a receptive area, with well-maintained properties, good land prices, adequate utilities, appropriate zoning, and potential development profitability compared with other sites and the possibility to introduce new roads or parking without significantly increasing traffic in the area (Fig. 10.4) (Farris 2001). Infill projects should adopt the architectural idiom of surrounding buildings to garner support from the community and create a cohesive design. Successful affordable infill housing projects must reconcile reducing costs through increased density while maintaining an attractive, desirable design.

The materials, style, and color of surrounding buildings should be incorporated into new developments following a study of existing patterns as was the case in Lethbridge, in Alberta, Canada by the author (Fig. 10.5). In Lethbridge, special attention was paid to building and planning details to ensure the proposed changes would achieve an urban and architectural fit. The designers were particularly concerned with roof shape and height, entrances and doors, window style, brick style, and color and an overall pattern and rhythm in the town's streetscapes.

When possible, infill housing should be placed facing existing roads and take their address from those streets. When new streets must be constructed, the present street pattern should be extended to create a smooth transition between new and old while maintaining public pedestrian access to both. Similarly, infill projects should respect current lifestyle patterns rather than trying to reform them. The existing lot sizes should be acknowledged and their patterns respected. Innovative design can ensure that infill housing will complement the neighborhood and will not disrupt the existing quality of life.

Infill projects should be created in areas close to relevant services and accessible transit. This is particularly important for elderly people, who require a specialized set of amenities close to the home. All residents would benefit from infill housing that capitalizes on public transportation routes already in place and that connects to existing infrastructure and circulation arteries. Again, mixed-use development is

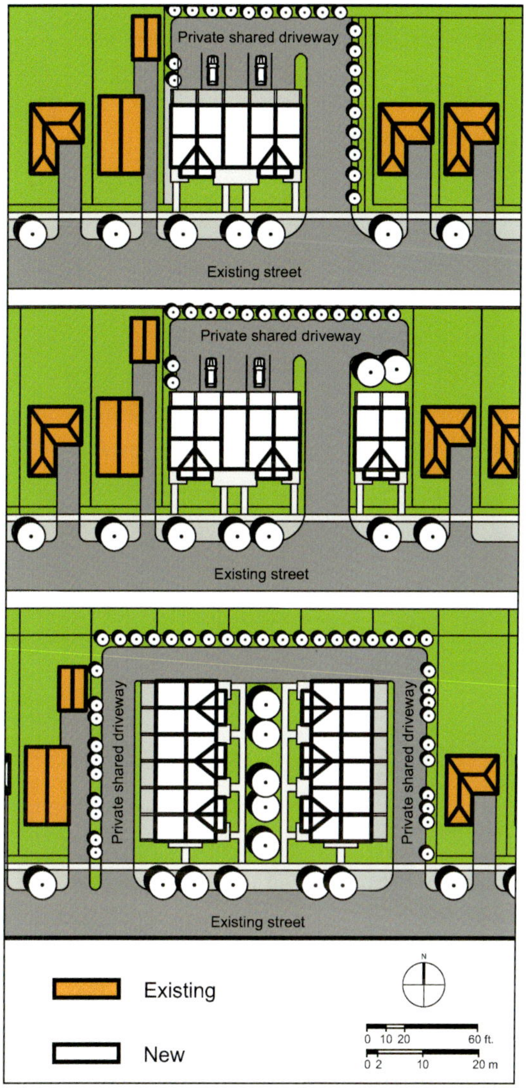

Fig. 10.4 An infill project needs to introduce new parking alternatives without significantly changing existing patterns in the area

important. By integrating residential and commercial spaces, infill developments can generate income that offsets the initial real estate and construction investment.

For sites with a large amount of housing, the type and cost of dwellings in the area should be varied. When possible, affordable housing should be introduced into a community at a minimum ration of one-in-ten, with one unit of affordable housing provided for every ten units of regular market housing. However, the socioeconomic profile of the area should accommodate the type of anticipated users. Proposing a project for low-income users in an area with costly services and amenities will not be successful.

Fig. 10.5 In Lethbridge, Alberta, Canada, special attention was paid to document the existing urban planning and building details to ensure that the proposed changes would achieve an urban and architectural fit

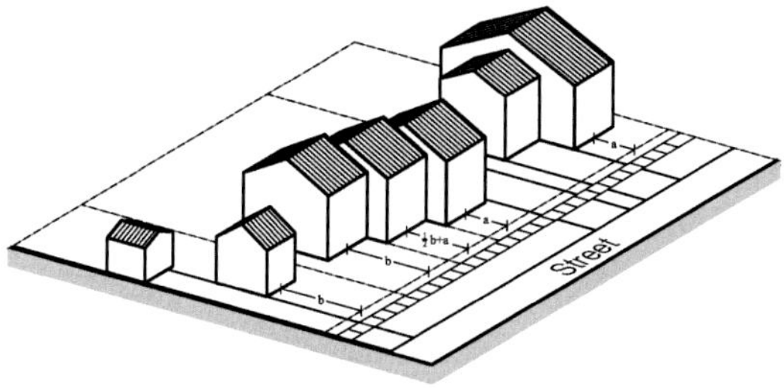

Fig. 10.6 Prior to design, planners, and architects should assess the setbacks of existing structures from roads to help determine the location of the new insertions

Site consideration is an essential part of the infill development process. Planners should conduct an assessment of the area which includes the physical condition of the site, its size, zoning parameters, and the presence and condition of existing infrastructure including the setbacks of existing structures from roads to help determine the location of new insertions (Fig. 10.6). In brownfield sites, remnants of the industry can be repurposed. Finally, a site can be selected in which dwellings can be oriented for passive solar gain, which will result in environmental and economic benefits (Friedman 2005).

There are significant benefits associated with infill development that will improve many dimensions of urban life, including economic, social, and environmental sustainability. Planners should use public–private partnership tools and minimize the transaction cost for developers to get infill projects completed (Farris 2001). Strategic partnerships and thoughtful site selection can result in infill projects that avoid the potential challenges and negative outcomes, and instead make neighborhoods more vibrant, diverse, and sustainable.

References

Carswell A (2012) The encyclopedia of housing, 2nd edn. SAGE, Los Angeles

Farris J (2001) The barriers to using infill development to achieve smart growth. Housing Policy Debate 1:1–30

Friedman A (2005) Homes within reach. Wiley, Hoboken

Kim J (2016) Achieving mixed income communities through infill? The effects of infill housing on neighborhood income diversity. J Urban Aff 38(2):280–297

McConnell V, Wiley K (2010) Infill development: perspectives and evidence from economics and planning. Resour Future 10:13

Squires G (ed) (2002) Urban sprawl: causes, consequences, and policy responses. The Urban Institute Press, Washington, DC

Wells W (ed) (2007) Blueprint for greening affordable housing. Island Press, Washington, DC

Chapter 11
Urban Design for Growth and Change

Abstract Adaptability to emerging realities is a key component of sustainable design. Flexible planning practices are a way to rise to the challenges presented by the ever-changing technological, social, and economic realities of today's urban environments. An adaptable approach to city planning based on long-term goals may also reduce energy and material consumption. This chapter explores urban design methods that foresee change and prepare for them to create resilient environments.

Keywords Flexible planning · Sustainable planning · Smart growth · Urban growth boundary · Connectivity · Pedestrian access · Density · Land-use planning · Mixed-use projects · Adaptable communities · Flexible urban outgrowth · Hierarchical importance · Greenfield development

11.1 Evolving Cities and Buildings

Historically, cities grow organically over centuries. When laws and codes governing planning and construction were not in place, dwellers, over generations, kept adding to their homes (Fig. 11.1). Yet, nowadays cities and buildings must be planned to adapt to and accommodate different types of change. The pressures of climate change, population growth, and economic instability force planners to think creatively about the future of urban communities. It is important never to assume that urban form that is responsive in the present will be sufficient for future conditions. Instead, planners must work to accommodate constantly shifting demands through flexible planning and building strategies.

Currently, 55% of the world's population lives in urban areas, a figure that is expected to grow to 68% by 2050. Rural-urban migration, along with overall population growth, could result in an additional 2.5 billion people living in cities worldwide over the next three decades. This immense demographic shift is accompanied by technological advancements, improvements in living standards, and climate change that may render certain regions of the world unlivable (United Nations 2018). Cities must respond to these trends through adaptable, sustainable planning.

A. Friedman, *Fundamentals of Sustainable Urban Design*,
https://doi.org/10.1007/978-3-030-60865-1_11

Fig. 11.1 Historically, cities grow organically over centuries. Dwellers kept adding to their homes or build new ones as was the case in Matera, Italy

The same adaptability must be extended to dwelling typologies. First-time buyers are increasingly challenged by expensive real estate markets, while changes in family structure, demographics, and way of life have reduced the need for outsized single-family homes. Instead, novel design and production methods must be taken into consideration to accommodate new markets of potential homeowners.

11.2 Planning Strategies for Growth, Change and Resiliency

Smart growth is a planning strategy that promotes communities not only as places to live, but as vehicles to promote health and well-being (Geller 2011). It originated as a method to reframe the policy debate regarding urban sprawl, but it has evolved into a method of adaptable planning that directly links the environment, the economy, and day-to-day life (Tregoning et al. 2002). Smart growth creates adaptable, resilient communities by protecting open spaces, revitalizing neighborhoods, and upgrading infrastructure to make it more efficient as was the case in the Benny Farm project in Montreal (mentioned in Chap. 9) where a community underwent densification (Fig. 11.2) (Geller 2011).

Throughout the twentieth century, urban development responded to demographic change through outward expansion. This pattern resulted in one of the central problems associated with sprawl, its erosion of the natural environment and impingement

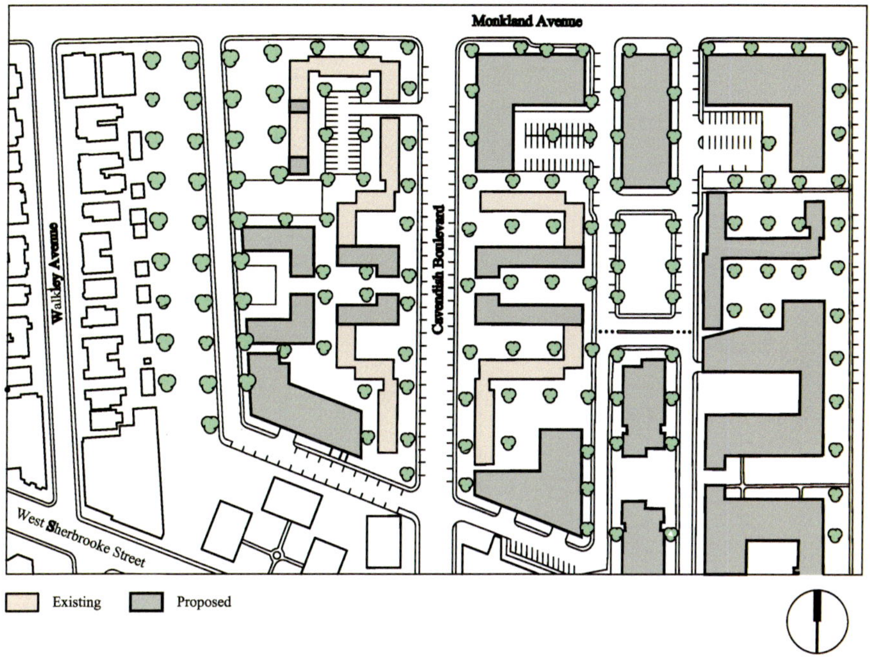

Fig. 11.2 Smart growth creates adaptable, resilient communities by revitalizing neighborhoods as was the case in the Benny Farm project in Montreal, Quebec, Canada that underwent densification

upon agricultural land. Smart growth places sustainability at the forefront of the development process, meaning creative solutions must be in place to preserve open spaces (Tregoning et al. 2002). One widely recognized example of a smart growth strategy is the urban growth boundary (UGB) in Portland, Oregon, USA. Since 1979, all land outside the UGB has been designated for resource use, thereby prohibiting outward urban expansion. Although this strategy has resulted in inflated housing costs, it has also been recognized to contain urban sprawl and protect natural resources and open space. Since the development of the UGB, Portland's neighborhoods have improved in terms of connectivity, pedestrian access, and density. This demonstrates that the city has succeeded in preserving open spaces, while allowing neighborhoods to adapt to the needs of a growing population (Song and Knaap 2004).

An essential component of smart growth is reinvestment in existing communities, including an integration of the new with the old. This includes infill developments in old neighborhoods as was the case in Cité Jardin Fonteneau in Montreal, Canada (Fig. 11.3). These developments must adequately address a site's context by adopting elements such as roof form, building typology, and masonry detailing. The architect's job is not merely to replicate these details, but to pay homage to them in a way that respectfully integrates the new with the old. Similar to the infill building process that was discussed in Chap. 10, the first step in the reinvestment process is a thorough documentation of the site context, including a visual and historical analysis, an

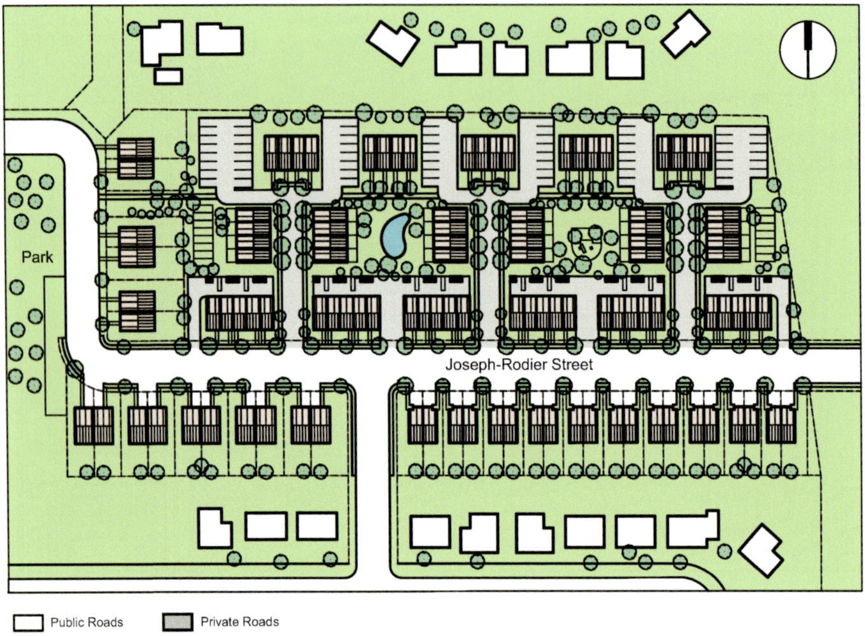

Fig. 11.3 An essential component of smart growth is the insertion of new infill projects in existing communities as was the case in Cité Jardin Fonteneau in Montreal, Quebec, Canada

understanding of current community demographics, an index of current amenities and businesses, and the zoning and land-use planning of the area. Herein lies one of the greates challenges associated with smart growth through community reinvestment: often, planners must take on enormous rezoning efforts to allow for mixed-use development and the autonomous local economy that comes with mixed-use projects (Friedman 2002).

Another key strategy for adaptable communities is the principle of hierarchical characterization. This is an analytical methodology that ranks each element of a city or suburb in order of importance (Fig. 11.4). Hierarchical categorization can guide growth by outlining the components of a development, their sequence of importance, and the relationship of components to each other. This framework can facilitate flexible urban outgrowth and infill projects. Early in the development process, planners agree upon a vision of the community they want to build. The vision at this stage is general, to allow for alternatives in built form, but definitive enough to exclude potential components that would alter the community's intended character. Next, planners articulate the key elements of design, including the major roads traversing the area. At the next level of hierarchical importance is the size and number of subdivisions within the community. In this stage, the location of community institutions and greenbelts are determined. The goal of hierarchical characterization is to define a broad vision of a neighborhood concept, which can be supplemented with design suggestions to create a harmonious aesthetic environment if development occurs in different stages (Friedman 2002).

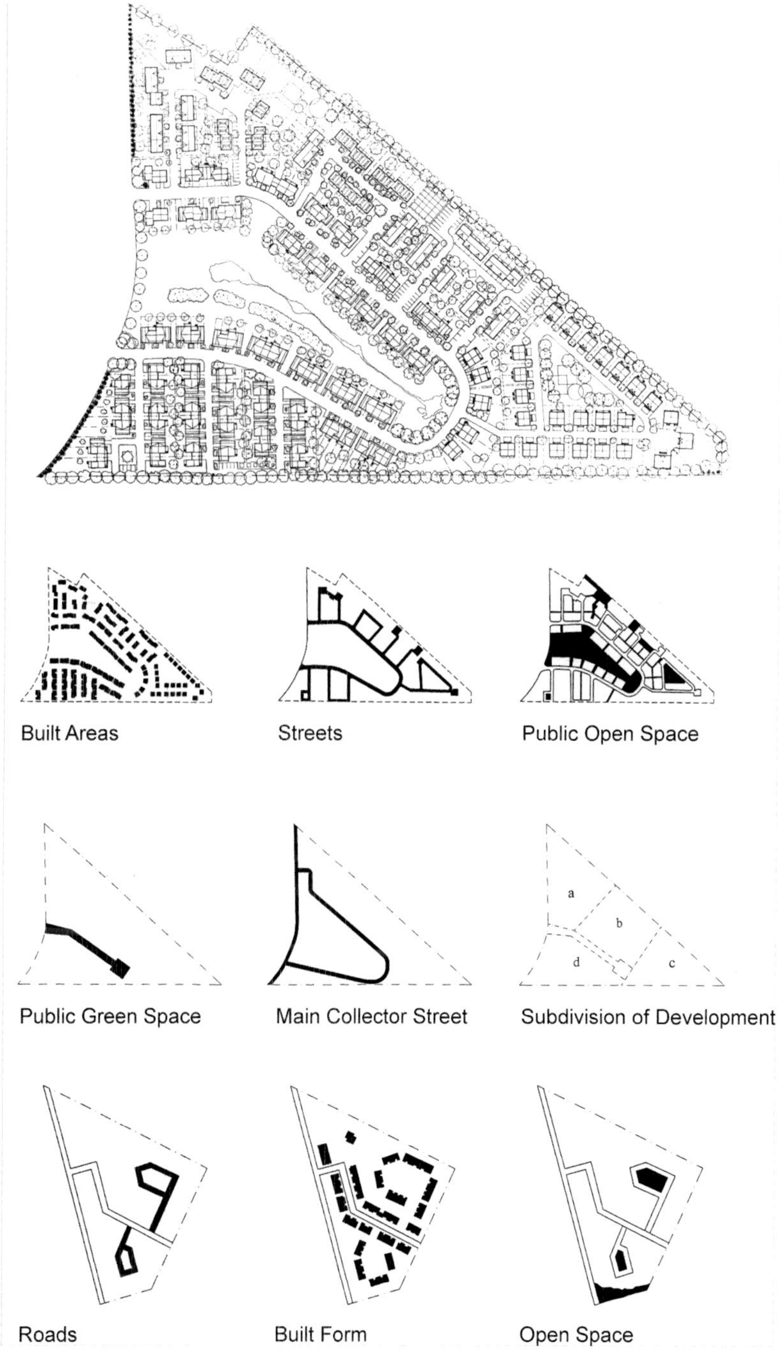

Fig. 11.4 Conventionally detailed planned community (top and second rows), a flexible approach to planning which divides the land to smaller segments (second row from bottom) and design of segment

Hierarchical characterization is particularly important in planning suburban communities on greenfield land. The community's potential to adapt to changes in the natural and social environment has important repercussions for its future viability. Before construction on previously vacant land can begin, components must be accepted at the regional or municipal level according to specific local requirements. Although specific requirements vary, the process of approval is generally quite lengthy, meaning that developers are forced to either speculate as to what amenities and housing types will be needed once the project is approved, or to design for present demands, which are constantly shifting. By refining the standard practice to create a hierarchical fragmentation of the development procedure, architects and planners would be able to design relevant housing for immediate demand. Hierarchical categorization in the development process will benefit all parties involved. The developer will build for market demand, the municipality will adjust development to its current priorities, and consumers will benefit from housing and amenities that meet their present needs.

In greenfield development, building and design codes should be developed to ensure a suitable mix of dwelling types and amenities, that meet the dynamic needs of residents, without sacrificing a harmonious environment. This can be accomplished through the use of a primary design code that is more descriptive than specific that may discourage against the construction of high-rise buildings without providing a specific height limitation. The primary design code should be supplemented with a secondary code that establishes dimensional and typological requirements, to avoid dramatically different interpretations of the primary code by competing developers (Fig. 11.5). The codes must be carefully constructed to allow for individual freedom of expression and adaptability, while ensuring that the character of the community is preserved.

11.3 Planning for Adaptable Buildings

Rapidly changing demographics and ways of life necessitate novel solutions. Increasing numbers of nontraditional households, an aging population, and technology that allows employees to work from home have created a demand for adaptable and expandable design methods.

Adaptable homes must be flexible to the changing needs of the homeowner. A structure that requires little to no internal load-bearing support creates a large, open floor space that maximizes adaptability. Multipurpose spaces within the open floor plan should be as large and square as possible, with a room size between 13.6 and 21 m^2 (44.6–68.9 sq. ft). Doors should be placed strategically within the space to allow it to be divided in two; for instance, placing a door in the corner permits another door to be added easily where necessary (Friedman 2013).

In addition to a flexible layout, circulation is an important consideration. Paths should be treated as multifunctional spaces to increase their efficiency and their ultimate adaptability. Halls can be widened to incorporate features such as tempo-

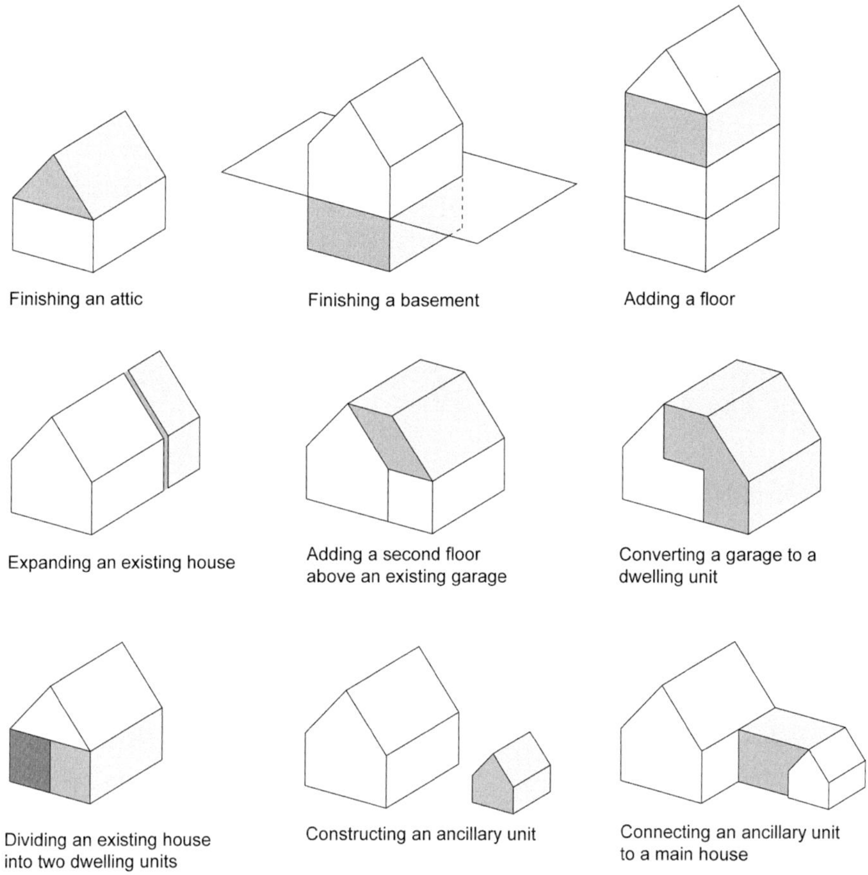

Fig. 11.5 The primary design code should be supplemented with a secondary form-based codes shown here for expending a dwelling unit that establishes dimensional and typological requirements, to avoid dramatically different interpretations of the primary code by competing developers

rary storage along the walls, which can be removed and repurposed if an extension requiring a staircase is constructed in the future. Residual spaces, such as alcoves in bay windows or under stairs, are good locations for storage and can be repurposed for future needs. It is essential that plumbing, ventilation, and heating systems are easily accessible, so they can be updated or moved if necessary. These services should include a vertical service shaft with accessible chases. Chases are horizontal tubes that contain the essential utilities needed for each floor of the dwelling (Fig. 11.6).

Expandable dwellings are an important type of adaptable design that allow occupants to increase their living space over time (Fig. 11.7). This allows homeowners to reside in the same space for longer and build a stronger sense of community with their neighbors. There are two main methods of developing expandable homes:

Fig. 11.6 It is essential that plumbing, ventilation, and heating systems are easily accessible, so they can be modified, like the chases in this illustration that contains electrical wiring

Fig. 11.7 Expandable dwellings are an important type of adaptable design that allow occupants to increase their living space over time as is the case in this Israeli apartment building where a balcony can be closed to add a room

"add-on" and "add-in" projects. Add-on designs involve expansion through addition. Horizontal expansion is only possible when homeowners are not seriously constrained by lot size, and developers must take into consideration the preservation of outdoor space when promoting these projects. Vertical add-ons are also possible by adding an additional floor above or below the original dwelling. The success of the add-on project depends on the versatility of the original design. It is important to use wall heights and materials that are easily expandable, as well as using the previously discussed techniques regarding layout and circulation.

The add-in method involves constructing a larger, partially complete shell to allow for later growth. In the first phase, only the primary living spaces are constructed, leaving open the opportunity for later development by the homeowners themselves to construct the internal space as they see fit. In this method, "wet functions" (e.g. bathrooms and kitchens) should be located centrally so that later adaptations do not require alterations to plumbing lines and fixtures. It is also important to initially plan for the later functions the home will require, such as central air conditioning and expanded plumbing lines (Friedman 2013).

New techniques of flexible home construction make necessary new forms of living possible. As technology allows more people to work from home, adaptable dwellings can facilitate an enjoyable live-work arrangement, and office space can be repurposed over time as one's career and family situation changes. As the global population ages, adaptable dwellings can allow for aging in place and multigenerational living arrangements that were not made possible by the traditional single-family home. Finally, clusters of small homes designed for potential add-on development allow young homebuyers to start with a small space and grow outwards as they start a family.

Adaptable building techniques have important implications for sustainability, on both the level of the community and the home. Preserving open spaces, reducing urban sprawl, promoting infill projects, and encouraging flexible home design can lessen the negative economic and environmental consequences associated with urban design that does not adequately prepare itself for future demands.

References

Friedman A (2002) Planning the new suburbia: flexibility by design. UBC Press, Vancouver

Friedman A (2013) Innovative houses: concepts for sustainable living. Laurence King, London

Geller A (2011) Smart growth: a prescription for liveable cities. Am J Public Health 93:1410–1415

Song Y, Knaap G (2004) Measuring urban form: is Portland winning the war on sprawl? J Am Plan Assoc 70(2):210–225

Tregoning H, Agyeman J, Shenot C (2002) Sprawl, smart growth, and sustainability. Local Environ 7(4):341–347

United Nations (2018) Department of economic and social affairs. https://www.un.org/development/desa/en/news/population/2018-revision-of-world-urbanization-prospects.html. Accessed 3 March 2020

Chapter 12
Identity and Diversity of Districts and Buildings

Abstract Urban and architectural coherence is associated with livable, successful cities and should therefore be considered by urban designers. Often, places gain coherence when their designers and builders follow a framework which was established years earlier. The intention is to avoid uniformity of appearance while creating a built environment where parts of the city will have their own distinct charm and identity to form an urban quilt of sort. This chapter looks at methods that foster such processes.

Keywords Macro-level coherence · Street design · Green spaces · Public places · Architectural idioms · Harmony and diversity · Visual order · Spatial quality · Cultural preservation · Heritage · Lot size · Setback · Parking · Building heights · Form-based codes · Mixed-use · Walkable

12.1 Planning for Urban Coherence

In general, variety of elements work together to create macro-level coherence in an urban environment. Street design, green spaces, public places, and architectural idioms work together to create a harmonious, complex atmosphere as it is often witnessed in old settlements like Positano, Italy (Fig. 12.1). To achieve a coherent urban environment at the macro level, there must be connectivity and relations between urban and architectural features of various scales (Salingaros 2000).

Urban design follows the same general rules that govern any complex system. In his article *Complexity and Urban Coherence*, Nikos Salingaros compiled a list of eight system rules that are essential to the study of urban design (2000); these rules detail the delicate balance of design elements that ensure a symmetric, but not homogenous, urban system (Salingaros 2000).

While Salingaros (2000) proposes a specific set of rules substantiated by visual, scientific, and urban evidence, the principles that underwrite his work are evident in other planning strategies for urban coherence. For instance, the idea that urban coherence is based on principles observed in the natural world is echoed through case studies of three planned neighborhoods in Rotterdam, the Netherlands. The

A. Friedman, *Fundamentals of Sustainable Urban Design*,
https://doi.org/10.1007/978-3-030-60865-1_12

Fig. 12.1 Street design, green spaces, public places, and architectural idioms that worked together to create a harmonious, complex atmosphere as it is often witnessed in old settlements like Positano, Italy

case studies also reiterate Salingaros' findings about the importance of small-scale variables such as window designs and proportions, roofs shapes, tree spacing, and scale of street lighting in creating urban coherence and the creation of guidelines at the macro level (Fig. 12.2). This definition of coherence emphasizes synchronicity between scales and a balance of harmony and diversity. The metrics of coherence are defined through "visual order" achieved through consistent, complementary arrangement of buildings, streets, and other design elements (Caliskan and Mashhoodi 2017).

In their 2017 study measuring the coherence of neighborhoods in Rotterdam, researchers Olgu Caliskan and Bardia Mashhoodi developed a model based on the Gini–Simpson index, which is a measure of the consistency of elements in a spatial system. The higher the value of the Gini–Simpson index achieved, the higher the level of coherence within the neighborhood. The research in Rotterdam revealed that the street block is the critical element in achieving urban coherence. The balance of design elements on a given street (e.g., architectural design, tree spacing, signage) is responsible for the observed spatial quality of the city (Caliskan and Mashhoodi 2017).

Traditional urban environments, composed of high-density, narrow streets are associated with higher levels of coherence as demonstrated in the city of Ypres, Belgium (Fig. 12.3). These neighborhoods are found to support greater diversity of services and shops, an essential component of urban vitality (Caliskan and Mashhoodi 2017). Planners must therefore take coherence into consideration and

	NOTABLE EXAMPLES	ILLUSTRATIONS	GUIDELINES
LOWER FLOOR OPENING	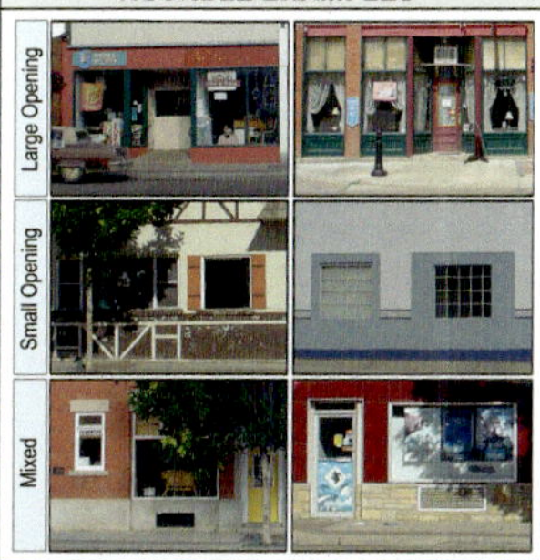	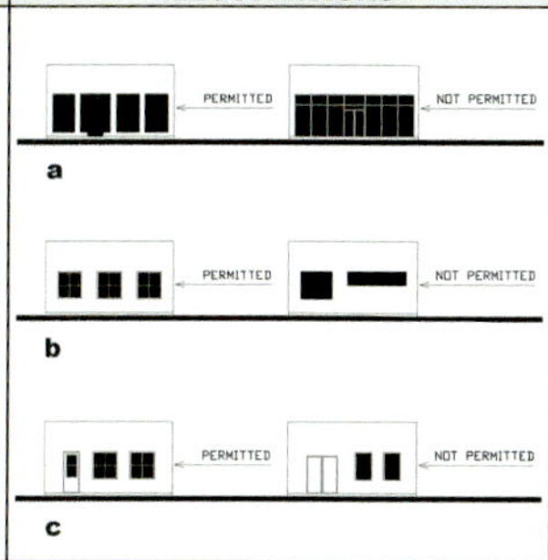	**a** Large openings of a commercial unit at lower floor should be divided into equal size divisions and respect the scale of their upper floor openings. No large curtain glass walls and reflective glasses are allowed. **b** Small openings of commercial units at lower floor should respect or repeat the scale and the rhythm of their upper floor openings. Big window should be divided and no narrow horizontal design is allowed as illustrated. **c** Entrance door frame should be at the same height as the upper frame of its adjacent windows. Panel and solid entrance door is not permitted on a commercial unit.
UPPER FLOOR OPENING		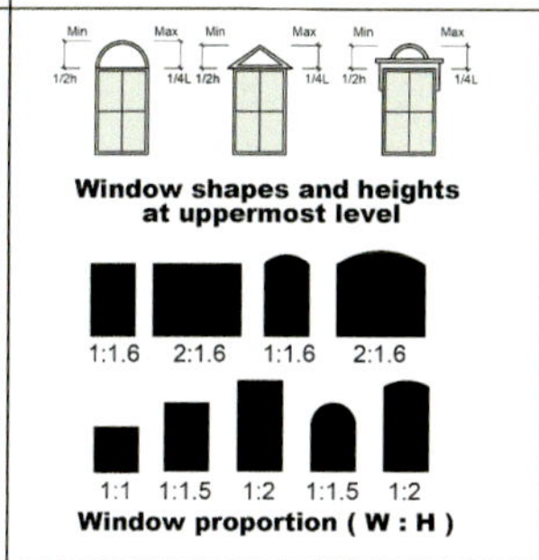	**Window Articulation** • All windows at the attic floor should have a decorative frame. • Frame shape and design shall correspond to neighboring buildings. • Decorative frame should not be taller than one-half of the height of the window or shorter than one-quarter of the height of the window. • All widows should be rectangular and oriented vertically. • Window proportion should respect the guildline pattern. • All front facades shall have no more than a 60% area of glazed opening or less than 30%. • No reflective glass is permitted.
ENTRANCE		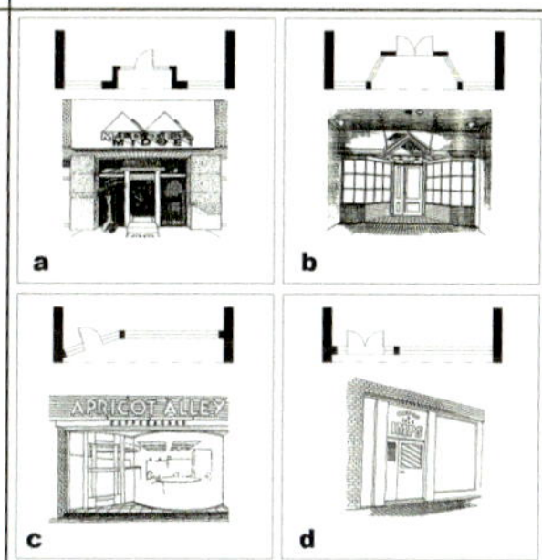	**a Recommended entrance design a:** Entrance located in the center of a modern style **b Recommended entrance design b:** Entrance located in the center of a traditional style **c Recommended entrance design c:** Entrance located off center of a modern style **d Recommended entrance design d:** Entrance located off center of a traditional style
ROOF AND PARAPET		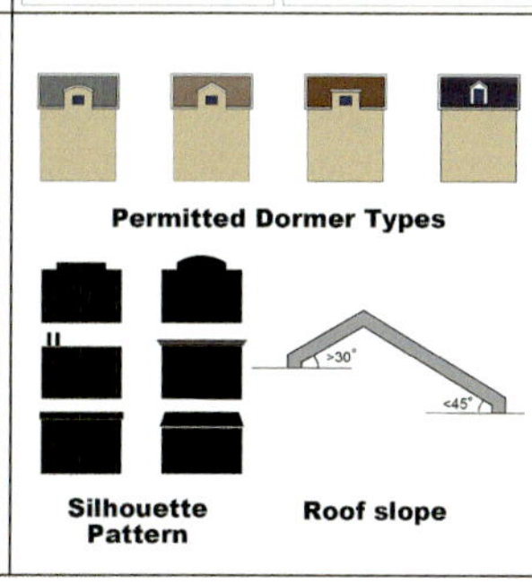	**Permitted Dormer Types** Four varieties of dormer are permitted at roof level. The design of dormer should respect the architectural context of downtown area. **Silhouette Pattern** Six varieties of parapet silhouette are permitted in the commercial district. The design of the cornice shall respect the local vocabulary and the neighboring buildings. **Roof Slope** 1.The slope of the roof addressing the street shall be greater than 30. 2.The slope of the roof addressing the rear of the unit shall not access 45.

Fig. 12.2 Small-scale variables such as window designs and proportions, roofs shapes, and scale of street lighting are vital to achieve urban coherence and create macro and micro levels guidelines

Fig. 12.3 Traditional high-density environments, composed of narrow streets are associated with urban coherence as demonstrated in the City of Ypres, Belgium

utilize tools such as Salingaros' eight system rules and Caliskan and Mashhoodi's adaptation of the Gini–Simpson index to achieve coherence in urban environments.

12.2 Creating Diversity, Identity, and Livability Within Homogeneity

As technology and ways of life change, so must the urban fabric of a city. However, cities and towns are an essential component of cultural preservation, and their historic design can serve as a gateway for tourists and residents to honor past ways of life as is illustrated in Bergen, Norway (Fig. 12.4). To maintain a diverse, modern experience in an urban place without sacrificing its distinct identity, planners must look to guidelines established when the urban environment was originally built. Urban designers are tasked with maintaining a delicate balance between society's need for evolution and the importance of preserving historical styles to create and maintain a sense of place (Orbasli 2000).

Coherence is defined in opposition to spatial fragmentation; however, this does not mean uniformity is ideal (Caliskan and Mashhoodi 2017). While achieving coherence between the old and the new is a critical component of urban design in heritage sites, harmony should not be confused with homogeneity. Instead, planners are responsible for deciphering the architectural language of a place. This essential

Fig. 12.4 Cultural architectural preservation can serve as a gateway for tourists and residents to honor past ways of life as is illustrated in Bergen, Norway

information includes the styles, zones, housing lifecycles, and original dwelling types originally included in an area. Decoding the heritage site's history will allow designers to make appropriate choices about dwelling proportions, essential architectural elements, and which buildings should be restored first or left unchanged (Friedman 2007).

In an intensive urban infill study of the neighborhood of Le Village in the city of Cornwall, in Ontario, Canada, the author developed guidelines for successful integration of infill housing at both the community and individual unit levels through a systematic approach and analysis. Starting with establishing patterns for the community, the urban investigation focused on Le Village's urban morphology and community structure, the community facilities, the industrial buildings, parking, alleys, and the waterfront area.

From these observations, the study area was divided into eight zones based on the morphology of the lots and their function and importance within the community. These divisions allowed for much more accurate generalizations about lot typology and therefore permitted more appropriate guidelines to be formed. Analysis of each zone yielded the general character and urban pattern of the neighborhood, upon which foundation guidelines could be developed. By comparing and analyzing data such as land coverage and floor area ratio, the general condition of land use and configuration in Le Village was obtained. Then, this information was synthesized, and the character of each zone established, considering various urban features, including lot size and setback, parking, and building heights (Fig. 12.5).

Existing urban conditions		Proposed Urban guidelines	
Lot	- 29.5' x 108.3' (9.0 m x 33.0 m) - 24.7' x 9.8' (7.5 m x 3.0 m)	Lot	- Side setback: 0.9 m (min.) on both sides
Parking	- Acess through the alley - Rear of the lot for parking - Curb parking (wide streets)	Parking	- Curb parking only (wide streets)
Land use	- Residential: duplex ,triplex, multiplex	Use	- Residential: single family, duplex
Grow space	- No alley wall - Spontaneous growth - No clear specified use	Grow space	- 1.5 (multi-use building) - Attached to a corner
Additions	- Additions do not respect the limit	Additions	- Respect the growth limit, street wall and additions limit
Sections and heights	- No backyard limit regulations	Sections and heights	- New trees in alley
		Corner buildings	- Buildings located on corners should relate the sides of the house to the street - Commercial use on corners is allowed

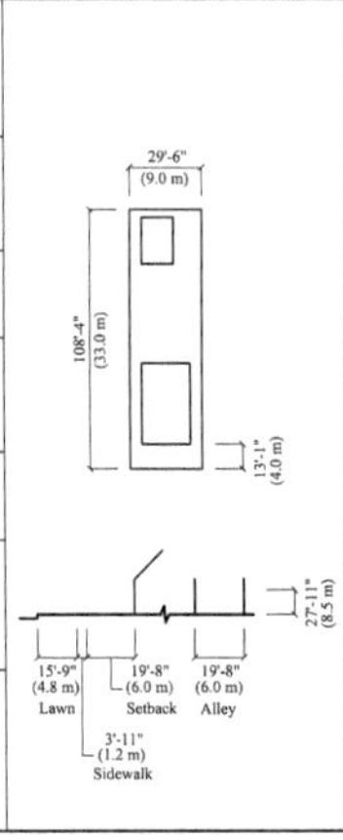

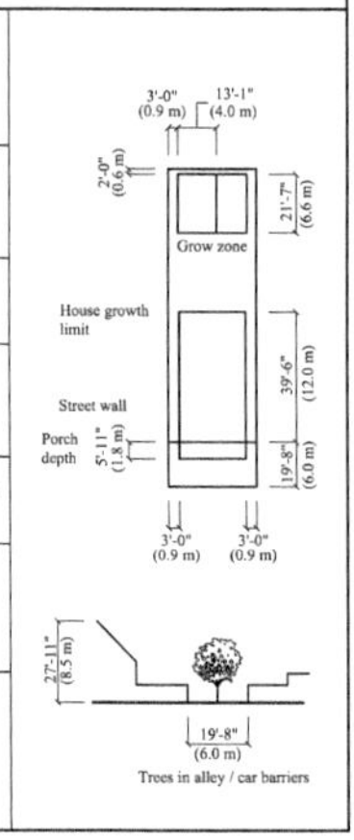

Fig. 12.5 Existing (left) and proposed urban guidelines (right) for one of the zones in Le Village, Cornwall, Ontario, Canada

Urban conservation efforts must enhance the community's livability while preserving the historic characteristics that give it a distinct identity (Orbasli 2000). Strewart Brand (1994) recommends that architects use an evolutionary design, with adaptations based on how a building originally looked and how it has changed over time. This same framework can provide guidelines to the designers of new infill homes. Elements such as cladding materials and color, decorative wood and brickwork, porch design, and outdoor features like trees, flower beds, and shrub species should be catalogued and developed into a formal code for developers.

12.3 Using Form-Based Codes

Form-based codes are a method of regulating land use through adherence to a specific urban form. They are designed to create better communities by requiring developers to adhere to specific design patterns, and they can be harnessed to create urban coherence, reduce sprawl, and ultimately improve communities through guidelines for the built environment. Form-based codes regulate the details that are most important for successful, livable neighborhoods such as building placement, height, and width; the design and layout of streets and blocks; and the details of the facades (Fig. 12.6) (Parolek et al. 2008).

Form-based codes are rooted in the developers' vision of the area. In infill developments, this vision is informed by the existing built environment and community (Parolek et al. 2008). Developing form-based codes in infill developments requires similar strategies to those previously discussed with heritage conservation. Designers must compile observations about existing dwellings involving elements

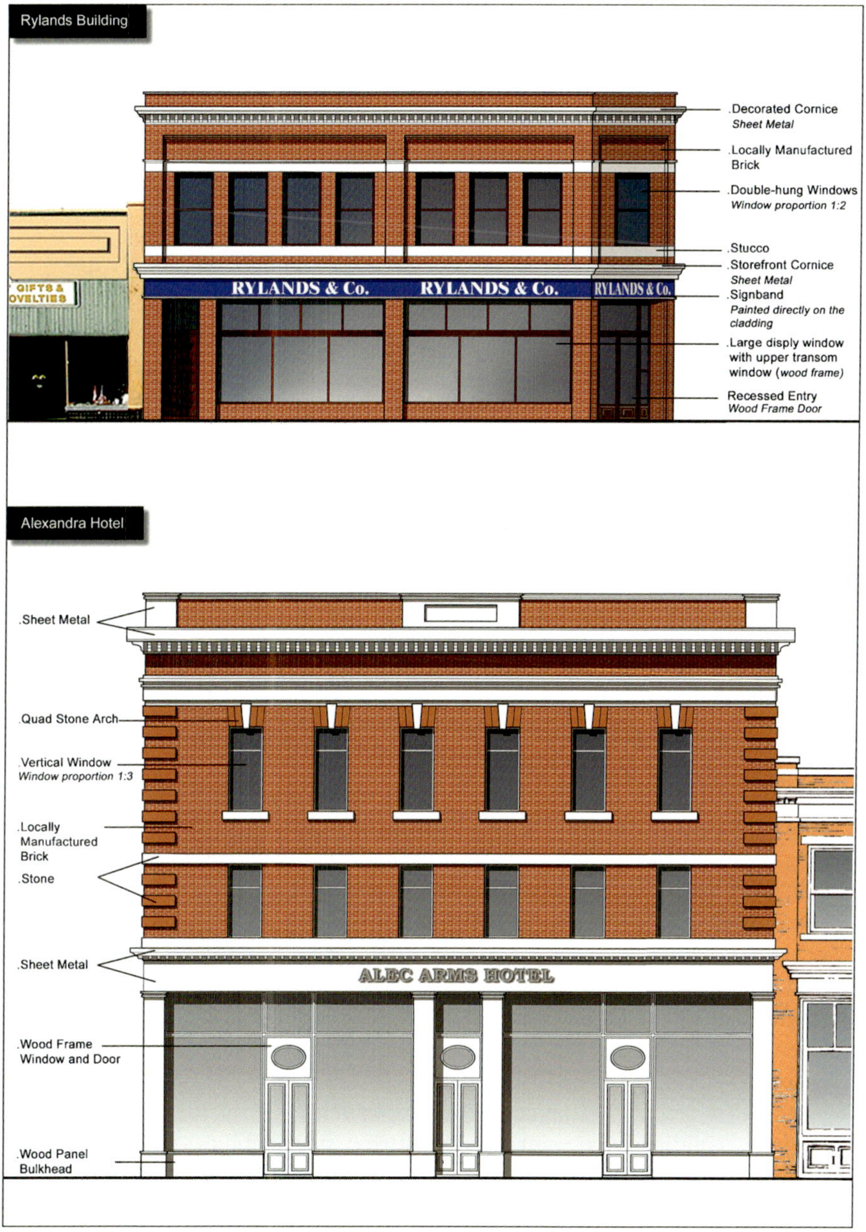

Fig. 12.6 Form-based codes for Lethbridge, Alberta, Canada, included features such as building placement, height, width, the design and layout of blocks, and facades' details

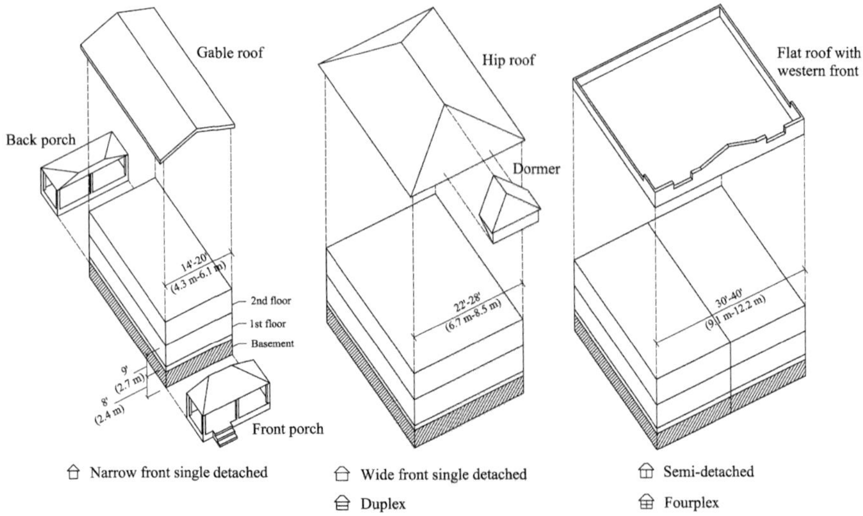

Fig. 12.7 Prior to the preparation of guidelines, designers can compile observations about existing dwellings involving elements such as cladding materials, porch type, windows, doors, roof designs and ancillary structures

such as building setbacks, cladding materials, porch type, windows, doors, and roof designs, ancillary structures, and landscape features (Fig. 12.7). These observations are then translated into formal guidelines. For example, if most porches in the neighborhood are made of wood or cement, the designer may include a guideline stating that new porches or porch additions must be constructed of either wood or cement to create continuity. In infill development, planners must prioritize the existing sense of place in a neighborhood through guidelines that allow for diversity without sacrificing the distinctive character of the community (Friedman 2007).

When developing a form-based code for a greenfield site, planners have a greater degree of control over their vision for the community. In this case, guidelines are derived from the developers' ideas about what the neighborhood should look like. The code includes similar features to those of infill developments; however, they are confined based on existing design elements. For instance, planners may restrict perpendicular or neon signs, or insist that all fences are white (Friedman 2002). Form-based codes for greenfield sites can be quite comprehensive, including standards for block type, building type, architectural styles, sustainability, and landscaping (Parolek et al. 2008). These standards detail requirements for street lighting, minimum and maximum building heights, the articulation of architectural elements, and other details that contribute to the identity of a neighborhood.

Form-based codes are an essential tool that allows for adaptable design that evolves over time while still preserving the historic character of an area. They allow planners to create and maintain mixed-use, walkable, diverse developments with a distinctive sense of place (Parolek et al. 2008). They allow the charm and local identity of a given area to be preserved through time and are a primary mechanism to achieving urban coherence.

References

Brand S (1994) How buildings learn: what happens after they're built. Viking, New York
Caliskan O, Mashhoodi B (2017) Urban coherence: a morphological definition. Urban Morphol 21(2):123–141
Friedman A (2002) Planning the new suburbia. UBC Press, Vancouver
Friedman A (2007) Sustainable residential development. McGraw Hill, New York
Orbasli A (2000) Tourists in historic towns: urban conservation and heritage management. E&FN Spon, New York
Parolek D, Parolek K, Crawford P (2008) Form based codes: a guide for planners, urban designers, municipalities, and developers. Wiley, Hoboken
Salingaros N (2000) Complexity and urban coherence. J Urban Des 5(3):291–316

Chapter 13
Planning for Energy Distribution and Waste Collection

Abstract While planning for passive solar gain was addressed in Chap. 6, this chapter discusses energy generation and its distribution in cities. Generation may involve the use of renewable energy like geothermal and solar sources; conservation may involve the positioning of trees and buildings to block prevailing winds; and distribution can involve the installation of district heating or cooling. Additionally, the chapter will recall methods of waste collection.

Keywords Renewable energy · Geothermal · Net-zero communities · District heating · Energy conservation · Local wind patterns · Sun orientation · Topography · Shadows · Wind turbines · Heat pumps · Waste collection · Recycling · Biomass · Biogas

13.1 Methods of Energy Generation in Urban Areas

In recent decades, the rate of environmental degradation has accelerated, and the conservation movement has become increasingly mainstream. To meet this trend, cities are reconfiguring and introducing new urban and architectural design strategies to achieve environmental goals. In recent years, policymakers and urban designers have focused on net-zero communities, cities that produce their own energy, and green forms of transportation, trends which will likely continue (Fig. 13.1).

A net-zero community, a relatively new concept, is one which has significantly reduced local energy requirements through efficiency gains and on-site power production by renewable energy (Fig. 13.2) (Carlisle et al. 2009). In 2014, the American Society of Heating, Refrigerating and Air-Conditioning Engineers published a report that detailed steps planners can take in transforming communities to achieve net-zero status. The process involves technical, economical, architectural, financial, legal, and behavioral components that should be studied by urban design professionals and adapted to the specific context in which the designer is working. Specific steps must be tailored to the local situation; however, the report suggests that some of the most important components for success in this process are bringing all stake-

A. Friedman, *Fundamentals of Sustainable Urban Design*,
https://doi.org/10.1007/978-3-030-60865-1_13

Fig. 13.1 In recent years, cities and designers adapted variety of green energy strategies including the introduction of solar panels over car parks, solar powered streetlights, and shared electric transport

holders together, establishing long-term goals, developing road maps, and managing the transition process (Zhivov et al. 2014). The management aspect is essential for planners; while the architectural tools included in net-zero communities are well-understood, no change can be accomplished without public support.

Fig. 13.2 A net-zero community, such as the ZHome in Issaquah Highlands, Washington State, US, by David Vandervort Architects, produces all its energy needs

Fig. 13.3 Large scale energy production using solar panels in Middlebury, Vermont, US, and wind turbines near Glasgow, Scotland

New developments can be specifically designed to produce their own energy. One of the most popular design methods in planning net-zero communities is utilizing alternative energy generators such as solar panels and wind turbines (Fig. 13.3). Solar panels can be placed on industrial buildings, to offset what would otherwise be significant emissions. Similarly, buildings can be equipped with wind turbines to produce energy close to where it will be used (Mertens 2002).

Designing for net-zero means changing how the city is used. One important point of change is how the city's residents move around. One alternative is electric cars, which play an important role in reducing human reliance on fossil fuels (Van Mierlo and Maggetto 2007). However, the optimal community is one in which the majority of people travel using active and public transportation (see Chap. 15). Therefore,

Fig. 13.4 The net-zero Sun Island development in the city of Almere, the Netherlands comprised of single-family homes fitted with solar panels

transit must also be designed to be net-zero. For this goal, the Netherlands leads the rest of the world. Since 2017, all electric Dutch trains have been 100% powered by wind energy, a model that other countries and cities will hopefully emulate (Agence France-Presse 2017).

The Netherlands has made significant strides in energy generation that extend beyond the train system. Almere Sun Island is a Dutch project designed to showcase the advantages of renewable energy. The Sun Island is a development in the city of Almere comprised of 520 solar panels, spanning an area of 7000 m^2. Annually, one-third of the district's electricity needs, and 10% of its energy needs, are met by the solar plant. Through the use of the district heating system (discussed below in Sect. 13.2) and solar energy, the residents' total carbon-dioxide emissions have been halved (Fig. 13.4). The community's success is amplified through individual efforts, including supplementary solar panels that residents have installed on their homes to further reduce the need for nonrenewable energy sources. Almere Sun Island is an example that other developments should try to emulate in the path to net-zero communities (Friedman 2017).

13.2 District Heating and Urban Design

A common source of heating and powering a district through a central source is referred to as district heating. The source can be of any kind, including solar, geothermal, or even fuel-based (Fig. 13.5). This type of system offers savings to each

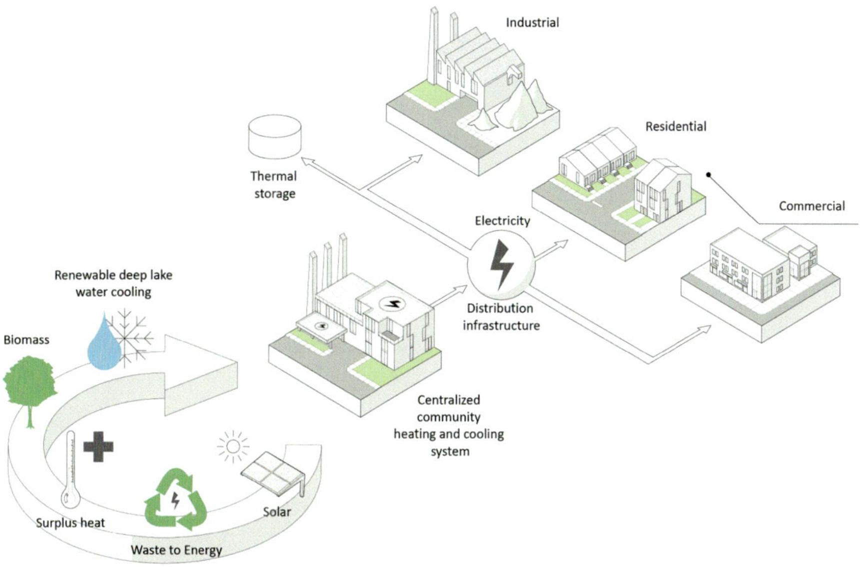

Fig. 13.5 A diagram showing a district heating system

household. In addition to developing a sustainable heating system, neighborhoods should be designed to maximize energy conservation by considering factors such as building size, orientation, windows, doors, and insulation.

Local wind patterns, sun orientation, topography, and shadows from other buildings and trees affect how buildings store and capture heat. Many of these features can be easily integrated at the community scale. For instance, placing houses along an east-west axis maximizes sunlight, while the selection of windows and doors improves insulation to conserve energy. Maximizing solar gain lowers energy consumption, which in turn lowers costs and reduces the emissions associated with producing energy through nonrenewable sources (Pacheco et al. 2012). To meet heating and cooling needs, district heating systems harvest renewable energy in a centralized plant. Water is collected in the plant and is either heated or cooled; it is later distributed to buildings within a neighborhood network. The elaborate system is made more efficient by utilizing renewable sources of energy. One option is geothermal energy, which is harnessed from the heat stored below the earth's crust. Geothermal energy is a natural resource that can be found almost anywhere, making it a promising option for district heating systems worldwide.

Solar power is another potential source of renewable energy, which is limited only by the complications associated with inconsistent sunlight throughout the year and across the globe. Wind turbines can be constructed in a community to convert wind into energy for human use (Fig. 13.6). All three options allow district heating systems to lower a community's collective environmental footprint by converting natural resources into heating and cooling services.

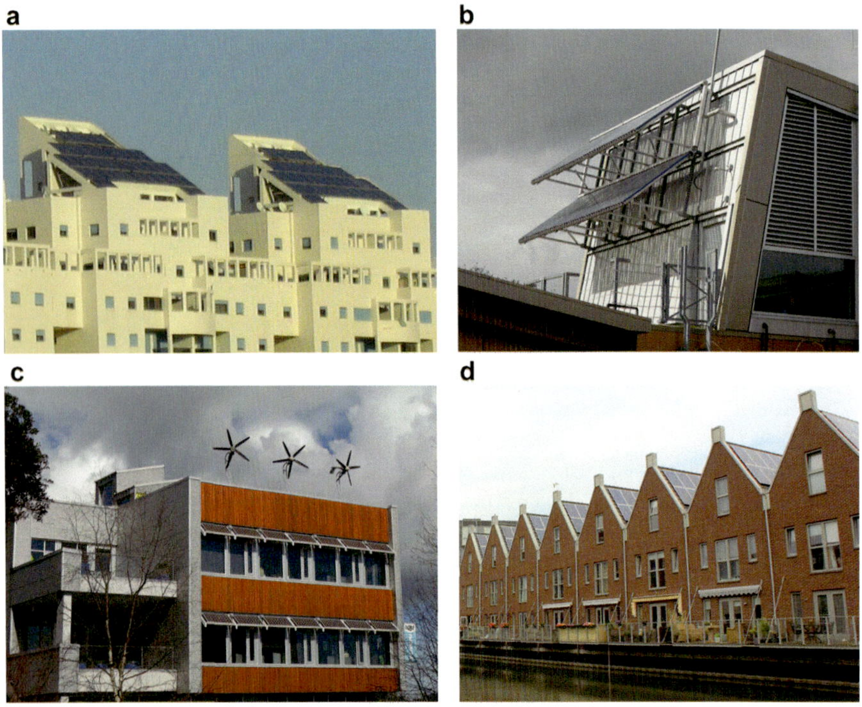

Fig. 13.6 Solar water heaters (top left), condescend tubs (top right), solar panels (bottom right) and small-scale wind turbines (bottom left) installed on buildings

Urban design both influences and is influenced by district heating systems. The system itself can make use of the existing natural landscape. For instance, in a neighborhood with a body of water, industrial heat pumps can extract heat from the lake, river, or sea, and transfer it to the central heating system. Features such as solar panels or wind turbines must be integrated into the city space and should be placed close to the central heating system and to buildings that consume high levels of energy.

In addition to lowering greenhouse gas emissions, district heating systems are low maintenance, have a long-life span, and eliminate the need for individual household heating systems. The space in homes that would otherwise be occupied by the individual heating system can be repurposed, thus increasing property values. Although district heating systems require a large initial financial investment, the switch to renewable energy will eventually result in cost savings for the district. Despite limitations and barriers to construction, district heating systems are an essential component of urban planning for efficient energy distribution (Friedman 2017).

13.3 Waste Collection and Recycling

Waste, despite its reputation, is a resource that can repurposed for energy generation. Biomass energy, the energy produced from low-carbon waste and plant materials, is a renewable resource. The process of incinerating biomass emits far less carbon than the comparable process of fossil fuel combustion, and it should therefore be considered as an option for waste recycling in communities (Wilkerson 2018).

Hammarby, a neighborhood in Stockholm, Sweden, is one of the world's leading sustainable communities. Hammarby's waste management system is one of its many innovative features. Pressurized air transports waste through pipes to central collection points, thereby eliminating the need for pick-up services and reducing the carbon produced in the waste collection process (Fig. 13.7). One-third of the waste at the plant is incinerated for use in community heating and cooling, while the other two-thirds are released as vapor water. Liquid sewage is converted into heat and biogas, which powers municipal buses. Solid sewage is used as compost for forested areas. This process means that the majority of the community's energy needs are met through renewable energy (Friedman 2017).

Dockside Green designed by Perkins and Will Architects, is an innovative community located on a former brownfield site near Victoria, British Columbia, Canada. The development has been rated LEED Platinum due to the application of a closed-loop design process. In other words, it is a self-sufficient and highly sustainable neighborhood.

Fig. 13.7 Hammarby's waste management system uses a pressurized air transports waste through pipes to central collection points, thereby eliminating the need for pick-up services and reducing the carbon produced in the waste collection process

Fig. 13.8 In the Dockside Green project in Victoria, British Columbia, Canada, water-usage is approximately 65 per cent less than a typical development, with water from treatment-plant on-site recycling used in toilets, soil irrigation and landscaping

As one would imagine, minimization of the community's environmental footprint is a result of many different efforts and processes. For one, much of the waste produced is later used to provide the development with power. Moreover, water-usage is approximately 65% lower than a typical development, with on-site water treatment, recycling gray water, soil irrigation, and landscaping (Fig. 13.8).

As part of the growing environmental movement, many committed individuals have pledged to live with zero waste. Even more ambitious is the movement for zero waste communities. Advocates for waste reduction propose a fourth "R" to the well-known "three Rs": reduce, reuse, recycle. The fourth "R" is redesign. Without a collaborative effort to redesign waste management infrastructure, zero waste is an impossible goal. The best way to accomplish this is on the community scale. Planners tasked with designing or redesigning a new neighborhood or subdivision should look to best practices in communities that have had success in reducing their waste and tailor those steps to the local context (Fig. 13.9). Selected steps in this

Fig. 13.9 Communal waste collection in the Skaftkärr project in Porvoo, Finland

process include improving recycling programs, introducing composting, and building reuse and repair facilities where people can bring appliances, furniture, and other items that they would otherwise put in the trash. Economic incentives, research facilities, and better industrial design can also contribute to reducing waste in a community (Connett 2013).

Starting from the level of the individual home, to the neighborhood, and ultimately to the city, planning for energy generation, distribution, and conservation requires collaboration among all stakeholders. Designers should look to successful projects, such as Almere Sun Island and Hammarby, to inform their final vision, while relying on the basic principles of building for energy conservation to implement this worthwhile goal.

References

Agence France-Presse (2017) Dutch electric trains become 100% powered by wind energy. https://www.theguardian.com/world/2017/jan/10/dutch-trains-100-percent-wind-powered-ns. Accessed 8 April 2020

Carlisle N, Van Geet O, Pless S (2009) Definition of a 'zero net energy' community. https://www.osti.gov/biblio/969716%20Carlisle%20et%20al.%202009. Accessed 6 April 2020

Connett P (2013) The zero waste solution: Untrashing the planet one community at a time. Chelsea Green Publishing, White River Junction

Friedman A (2017) Designing sustainable communities. Bloomsbury, London

Mertens S (2002) Wind energy in urban areas: concentrator effects for wind turbines close to buildings. Refocus 3(2):22–24

Pacheco R, Ordóñez J, Martínez G (2012) Energy efficient design of building: a review. Renew Sust Energ Rev 16(6):3559–3573

Van Mierlo J, Maggetto G (2007) Fuel cell or battery: electric cars are the future. Fuel Cells 7(2):165–173

Wilkerson J (2018) Biomass over coal: burning different carbon to mitigate climate change. http://sitn.hms.harvard.edu/flash/2018/biomass-over-coal-burning-different-carbon-to-mitigate-climate-change/. Accessed 18 May 2020

Zhivov A, Case M, Liesen R, Kimman J, Broers W (2014) Energy master planning towards net-zero communities and campuses. ASHRAE Trans 120(1):114–129

Chapter 14
Communities with a Digital Heart

Abstract In the development of any modern-day community, strong consideration must be given to digital infrastructure during the early design phase. The Internet has evolved significantly over the past decade to become a critical component of a community infrastructure. The number of connected devices is proving to be an exponentially growing trend globally. These devices include a vast array of sensors, monitoring devices, and intelligent controllers. Establishing the infrastructure to collect data from these sensors and respond in an intelligent way creates a community that is not only smart, but also allows analytics to be performed and insights to be developed for improvements in the future. This chapter looks at this trend to study its effect on urban design.

Keywords Sensors · Monitoring devices · Intelligent controllers · Digital infrastructure · Interconnected digital infrastructure · Smart city · Digital cities · Digital age · Car sharing programs · Wi-Fi signals · Healthcare data · Digital tools

14.1 The New Digital City

By some measures, many cities are already digital: residents communicate and navigate using smart phones, tourists access public Wi-Fi, and smart appliances have made their way outside the home and into the public (Fig. 14.1). Urban systems are now on their way to full connectivity through the "Internet of Things" (IoT). Cities that integrate interconnected digital infrastructure have the capacity to improve community safety, public health outcomes, transit data, and social cohesion. However, the new digital city comes with added risks associated with big data, notably issues of privacy (Elmaghraby and Losavio 2014). Urban designers are tasked with the challenge of integrating technology into cities in a way that is both adaptable and safe.

While digital cities may sound futuristic, projects that integrate technology into urban infrastructure have been underway for decades. In Toronto, Canada, the Sidewalk Toronto project that aimed to drive sustainable, inclusive urban development along the city's waterfront using digital technology and was recently was sus-

A. Friedman, *Fundamentals of Sustainable Urban Design*,
https://doi.org/10.1007/978-3-030-60865-1_14

Fig. 14.1 Numerous small and large cities offer free Wi-Fi service to citizens and visitors like Sintra, Portugal

pended due to financial considerations. Sidewalk Toronto was intended to demonstrate how integrating technology into urban planning can yield significant improvements to quality of life. Those who contributed to the plan (many of whom were ordinary Torontonians) felt strongly that technology should not be integrated unless it was a proven improvement to existing solutions. Sidewalk Toronto responded to this by integrating digital infrastructure with the plan's priorities of environmental sustainability and inclusivity: for example, ensuring that the community is powered by electric energy at a comparable price to current energy bills (Dawson 2018).

Projects like Sidewalk Toronto are made possible through the IoT, which integrate multiple systems to collect data and provide digital services. Urban IoT are based on the "Smart City" paradigm and are designed to improve municipal governance, public service provision, and quality of life for city dwellers (Fig. 14.2). This goal must be supported by political, financial, and technological infrastructure that allows for communication between different systems and easy access to data and

Fig. 14.2 Urban Internet of Things (IoT) are based on the "Smart City" paradigm and are designed to improve municipal governance, public service provision, and quality of life for city dwellers

services. To function, the IoT utilizes a web service architecture for both end users and developers; this approach is promoted by international standardization bodies and research projects. Due to the large service area of an urban IoT system, technologies must be used that can support a high volume of traffic. Finally, both end users and developers will require digital and physical devices to access the data and services provided by the IoT (Zanella et al. 2014). The significant investment of human and financial capital that is required to establish an urban IoT has the potential for meaningful returns. Smart cities can reduce environmental impacts, improve public services, and enable better use of public resources (Mohanty et al. 2016). The mechanisms behind these improvements will be discussed in greater detail in the following section.

Some of the most serious concerns about the feasibility of digital cities are those of security and privacy; however, just as IoT technology is under development, so are measures to protect citizens' digital data. The IoT enormous benefit is derived from the information it collects and generates. Data about people's health, activities, travel patterns, and preferences can make cities more convenient, efficient, accessible, and safe. Nonetheless, any use of big data raises legitimate concerns about information security and validity, as well as legal and social constructs relating to privacy (Elmaghraby and Losavio 2014). To address these concerns, planners, designers, and engineers must integrate sophisticated technologies into the urban IoT to block third party access to data and to decouple urban data collected in public spaces from private data (Bartoli et al. 2011).

14.2 The Use of Digital Tools in Urban Design

The novelty of urban IoT enables invention and creativity among planners and policy makers; a truly smart city requires innovative strategies for integrating digital tools in urban design. Many inefficiencies in today's cities can be identified and improved through technology.

In addition to the broad technical architecture that is required for urban IoT, cities require smaller systems and flows of information to maximize the benefits of the digital age.

The benefits of many existing technologies are already apparent in the urban context. The rise of online grocery shopping allows elderly or disabled residents to access fresh, healthy food without putting themselves at risk (Hand et al. 2009). Transit apps for smart phones have already revolutionized how people move around the city and improved people's experiences of taking public transit (Dutzik et al. 2013). Cultural institutions can use the Internet to share information with city residents about upcoming events or to gain a following for a smaller event (Fig. 14.3). Many museums have even begun to offer virtual tours, allowing people who cannot travel the privilege of important cultural experiences. Library cardholders can now access e-books and audiobooks. Information about all of these services, as well as public support services, can be disseminated by the city through newsletters, social media, or other digital sources.

Constructing a useful smart city system involves transferring technology from the level of the individual to the public. Many public transit users rely on digital technology to plan their commute, and the added convenience and assurance of up-to-date arrival times reduces the barriers that limit public transport usage (Dutzik et al. 2013). Using the IoT, technology associated with public transit or car sharing programs can also be harnessed to improve the service itself (Fig. 14.4). Collecting transport data using smart cars, cameras, and 4G systems can result in efficient route management and can assist in the central transit authority's decision-making (Hashem et al. 2016).

Smart home technology is similarly useful to the urban IoT project. Consumer electronics that allow individuals to monitor and control aspects of their home, including energy usage, can be transformed and applied to the level of the city (Fig. 14.5) (Zhou et al. 2016). The urban IoT can facilitate the use of a smart grid system, which enables researchers to utilize real-time power generation and consumption data. The system requires smart grids and smart readers to be placed throughout the city, which use Wi-Fi signals to track energy usage (Hashem et al. 2016). Smart grid systems are associated with improved energy efficiency. Moreover, the data collected through this system improves decision-making about current and future energy needs (Al Nuaimi et al. 2015).

Finally, cities can gain valuable insights based on aggregate health data to make informed public health decisions. Analytics of healthcare data can be used to predict epidemics, cures, and diseases, as well as improve quality of life and avoid preventable death (Hashem et al. 2016). Public health is an essential concern of

Fig. 14.3 Cultural institutions and cities can use the Internet to share information with city residents and visitors about upcoming events or notable landmarks such as the Truli in Alberbello, Italy

urban design; indeed, components of the built environment can have tremendous effects on people's mental and physical well-being (Jackson 2003). Cities can increase their capacity for preventative care by analyzing healthcare data, collected through wearable devices, to identify early markers of disease and disease clusters (Hashem et al. 2016).

Digital technologies create new and challenging opportunities in the field of sustainable urban design. The potential of urban IoT lies in how services can be connected and how valuable data can be extracted from technology to improve the experience of the city. Although 'smart cities' may sound like science fiction, technological innovation is already allowing planners to integrate digital tools into urban design for the improvement of cities' efficiency and their residents' quality of life.

Fig. 14.4 Using the IoT, technology associated with public transit or car sharing programs can be harnessed to improve the service

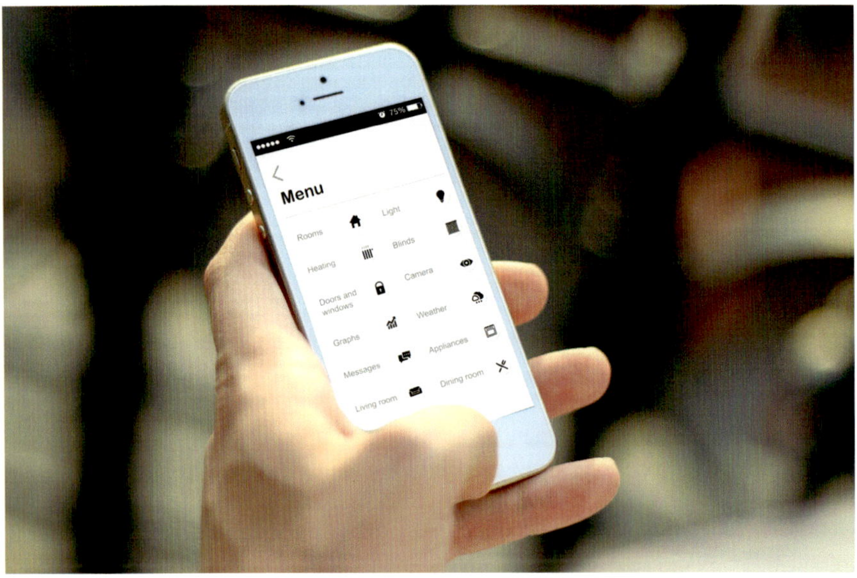

Fig. 14.5 Consumer electronics allows homeowners to monitor and control aspects of their home, including energy usage, safety and light control

References

Al Nuaimi E, Al Neyadi H, Mohamed N, Al-Jaroodi J (2015) Applications of big data to smart cities. J Internet Serv Appl 6(1):25

Bartoli A, Hernandez-Serrano J, Soriano M, Dohler M, Kountouris A, Barthel D (2011) Security and privacy in your smart city. Paper presented at proceedings of Barcelona smart cities congress, Barcelona, 29 Dec 2011

Dawson AH (2018) An update on data governance for sidewalk Toronto. Sidewalk Labs. https://www.sidewalklabs.com/blog/an-update-on-data-governance-for-sidewalk-toronto/#:~:text=Sidewalk%20Labs%20believes%20an%20independent,urban%20data%20originating%20in%20Quayside. Accessed 25 April 2020

Dutzik T, Madsen T, Baxandall P (2013) A new way to go: the transportation apps and vehicle-sharing tools that are giving Americans the freedom to drive less. https://uspirg.org/sites/pirg/files/reports/A%20New%20Way%20to%20Go%20vUS1_1.pdf. Accessed 22 April 2020

Elmaghraby A, Losavio M (2014) Cyber security challenges in smart cities: safety, security, and privacy. J Adv Res 5(4):491–497

Hand C, Riley F, Harris P, Singh J, Rettie R (2009) Online grocery shopping: the influence of situational factors. Eur J Market 43(9/10):1205–1219

Hashem I, Chang V, Anuar N, Adewole K, Yaqoob I, Gani A, Ahmed E, Chiroma H (2016) The role of big data in the smart city. International Journal of Information Management 36(5):748–758

Jackson L (2003) The relationship of urban design to human health and condition. Landsc Urban Plan 64(4):191–200

Mohanty S, Choppali U, Kougianos E (2016) Everything you wanted to know about smart cities: the Internet of Things is the backbone. IEEE Consum Electron Mag 5(3):60–70

Zanella A, Bui N, Castellani A, Vangelista L, Zorzi M (2014) Internet of Things for smart cities. IEEE Internet Things J 1(1):22–32

Zhou B, Li W, Chan K, Cao Y, Kuang Y, Liu X, Wang X (2016) Smart home energy management systems: concept, configurations, and scheduling strategies. Renew Sustain Energy Rev 61:30–40

Part III
Mobility and Connectivity

Chapter 15
Mobility and the City: The Broad View

Abstract Many of the trips taken by people and the mode of transit they use to reach them are linked to the land uses decision and the location of amenities. The siting of institutional, commercial, and residential buildings will affect those voyages. On a macro-level, attention must be given to where to locate them, cluster them perhaps, and create efficient and pleasant modes to reach them by foot, bikes, or public transit. This chapter examines those aspects as a basic underlying principle of urban design.

Keywords Active transportation · Higher densities · Land-use management · Locating amenities · Sustainable mobility · Walkability · Planned Unit Development (PUD) · Natural settings · Public health · Renewable energy sources · Green technologies · Low impact development · Green modes of transit

15.1 Higher Density and Mobility for Efficient Reach

In general, it is easier to adopt active transportation in areas with efficient transit networks and higher-density locations such as city centers. Density allows for a bigger volume of people to be located near certain amenities and is in turn an aspect which motivates the building of amenities as was the case in Ørestad, near Copenhagen, Denmark. It is Copenhagen's most recent expansion. Plans were developed in 1997 and building began in 2001. A linear-development plan was utilized, with focus placed on a central transport artery. This main artery consists of a light rail system, bicycle lanes, and a motorway. The light rail system has intentionally been raised above ground level. This prevents the system from acting as a barrier to pedestrians and cyclists, while also firmly asserting its presence in the community (Fig. 15.1).

Mixed land use is another important planning component which allows different amenities to be clustered in a single location. Although planning to support transport infrastructure may be challenging, there are strategies for such infrastructure to be successfully developed, including investments guided by land-use management (Cervero 2009).

A. Friedman, *Fundamentals of Sustainable Urban Design*,
https://doi.org/10.1007/978-3-030-60865-1_15

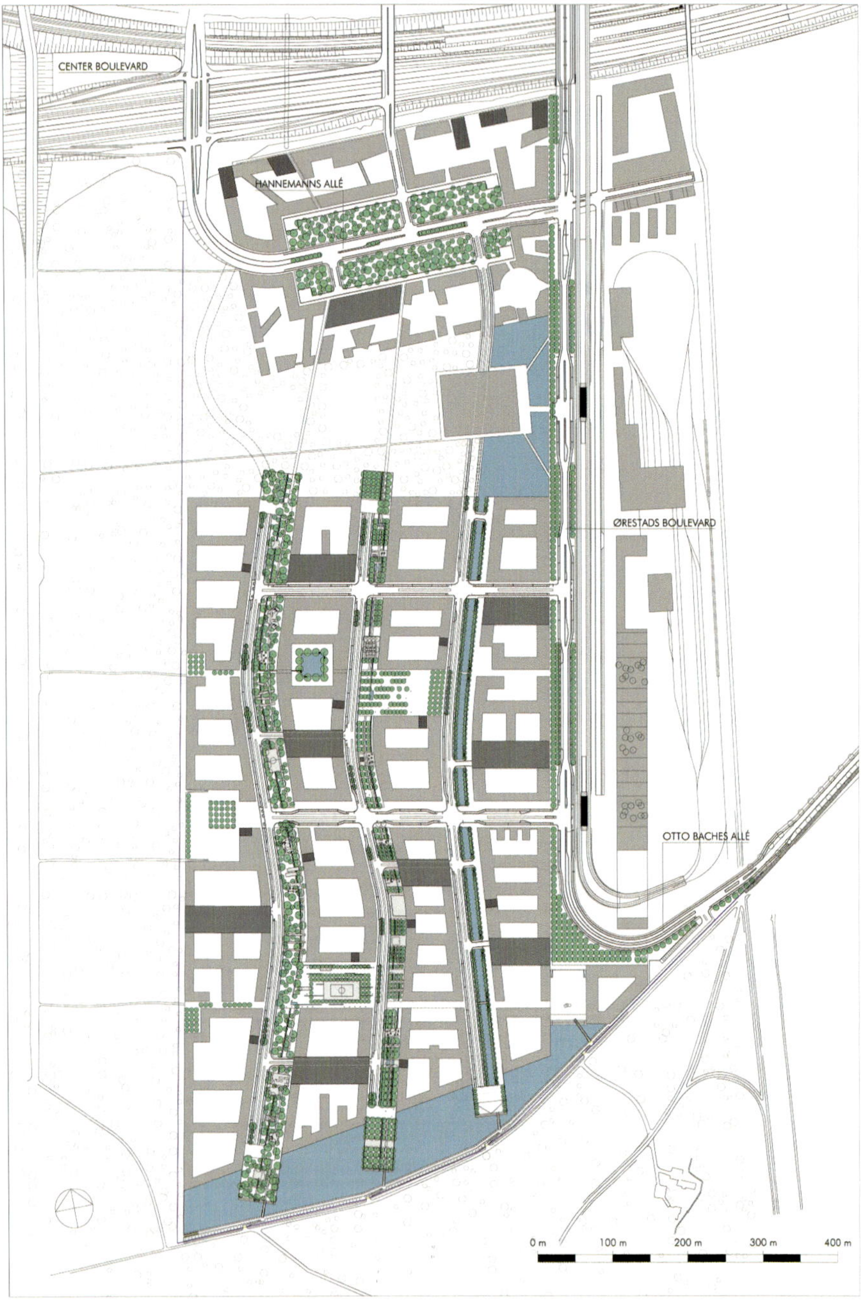

Fig. 15.1 A linear-development plan was utilized in Ørestad near Copenhagen, Denmark with focus placed on a central public transport artery that consists of a light rail system and bicycle lanes

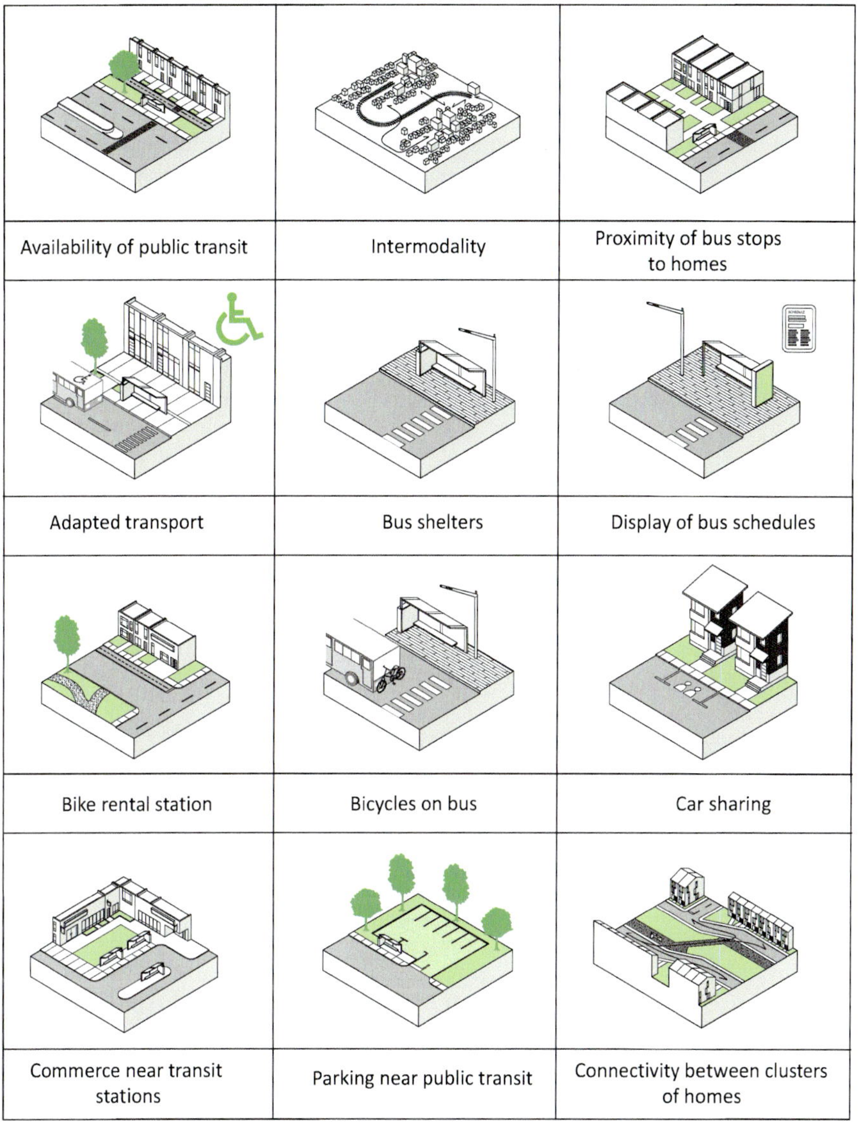

Fig. 15.2 Overarching planning considerations and strategies with effect on mobility in communities

Several overarching planning aspects of communities will affect mobility (Fig. 15.2). For example, locating amenities in a way which makes them accessible by means of active transportation such as walking, cycling and public transport is very important if the aim is to encourage the use of sustainable modes of transport. Amenities such as shops, companies, and zones where economic activities are present should be served by urban transport networks to provide good accessibility

(Commission of the European Committees 2007). Indeed, studies have found that land-use policies placing residents closer to destinations and offering them more alternative transportation options contributes to less car-dependence and to more walking.

Given that proximity plays a significant part in terms of incentivizing individuals to travel by means which are non-motorized, density should be prioritized first and foremost. Indeed, if public transit networks are to be self-sustaining, then they must serve a high enough volume of riders which may only be achieved if population densities are sufficiently high.

A macro land-use planning is another important aspect to address in promoting sustainable mobility; the aim is to design for comfortable walking distances (Fig. 15.3). For instance, placing schools at the center of communities may play an important role in enhancing walking and biking among children. Locating services

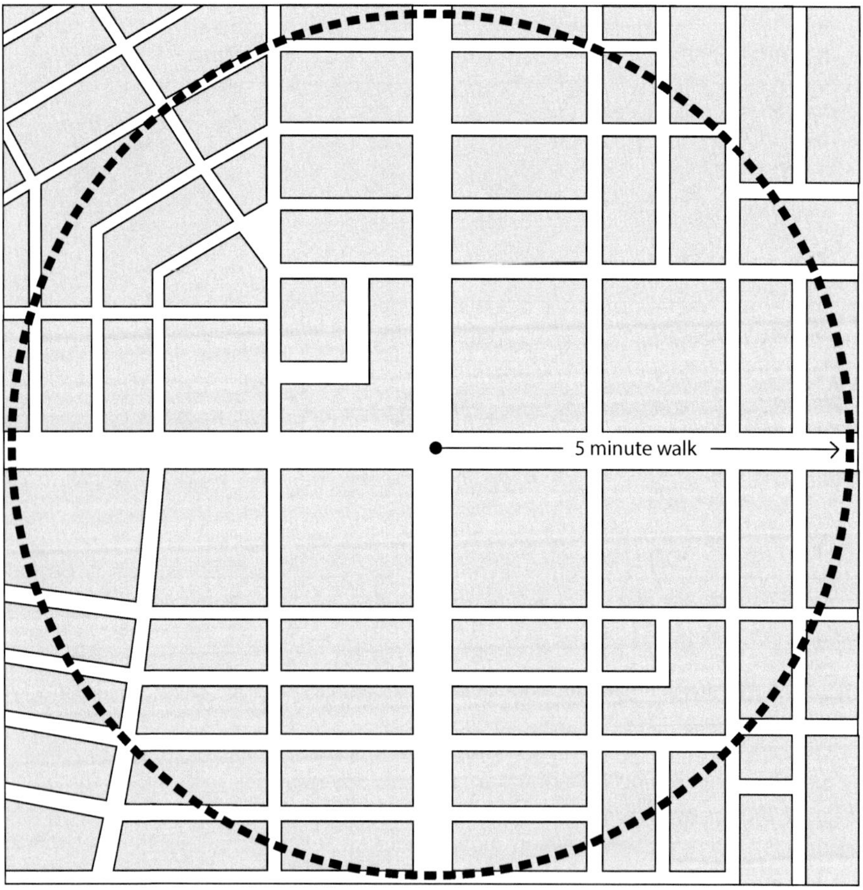

Fig. 15.3 People will reach amenities by foot when they are located within a comfortable walking distance from residences

and commercial amenities at the intersections of arterial roads which are in proximity to residential developments is another good strategy to promote walkability. Furthermore, new public transit networks should be designed while considering existing streets and paths in order to build on what is already there.

A challenge concerning locating amenities is that their place is often determined according to economic prospects, and number of users they will attract. However, proper land-use planning plays an important role in making certain amenities accessible which affects their economic value. Cost may indeed be determined by using a Planned Unit Development (PUD) design process which allows for more flexibility in making decisions about land use. For instance, planning in a way that promotes good pedestrian connectivity may indeed present positive economic outcomes (Cervero 2009).

An important urban design aspect is to balance economic productivity and community place-making by creating a new relationship between public infrastructure and their place in cities (Cervero 2009). Indeed, a common trend exists where planning for transportation is not undertaken solely for the purpose of achieving economic goals, but also to promote place-making and community building, which in turn makes places more sustainable as explained in Chap. 12 (Cervero 2009). The transit stops can be beautifully design and become community hubs as is the case in Ørestad, near Copenhagen (Fig. 15.4).

Fig. 15.4 The transit stops can be beautifully designed to become a visual attraction as is the case in Ørestad, near Copenhagen, Denmark

15.2 Green Modes of Transit and Active Mobility

As outlined in previous chapters, for development to be sustainable, planners must prioritize the needs of pedestrians over vehicles (Cervero 2009). This can be achieved by investing in active mobility, increasing its public appeal and minimizing car-centric infrastructure. Active mobility is when an individual uses their own power to travel; this can include walking, running, biking, skateboarding, rollerblading, and using a wheelchair (Government of Canada 2014).

A case in point is the community of Bois-Franc in Montréal, Canada, that was designed to promote an active lifestyle and utilize outdoor space. Through stringent urban planning techniques, the developers have managed an incredible feat in preserving the existing natural environment while creating a dynamic, livable space home to 2883 residential units. Several man-made features interact extensively with the preexisting environment, maximizing opportunities for residents to enjoy the outdoors. The community is lined with an extensive network of sidewalks and bike paths. This increases safety for pedestrians and bicyclists on the streets and demonstrates the developer's commitment to active modes of transportation. The cycling paths, running through streets and natural settings alike, serve well for both commuting and leisure purposes (Fig. 15.5).

Active mobility benefits individual and public health, the environment, and the economy. As the European Heart Network states, transition to active mobility and public transportation can only occur if sustainable transport choices become more attractive, convenient, and affordable than private vehicular transport (European Heart Network 2008). There are existing strategies that demonstrate how mobility, livability, and sustainability can be achieved and promoted (Cervero 2009). Among them include multiway boulevards and greenways connected to active mobility networks and well integrated TOD (Fig. 15.6) (Cervero 2009). Planners should design transit systems to be sustainable in every way possible, using renewable energy sources, green technologies and low impact development.

Fig. 15.5 Bois-Franc, near Montreal, Quebec, Canada was designed to promote an active lifestyle and utilize outdoor space by including many pedestrian bike paths

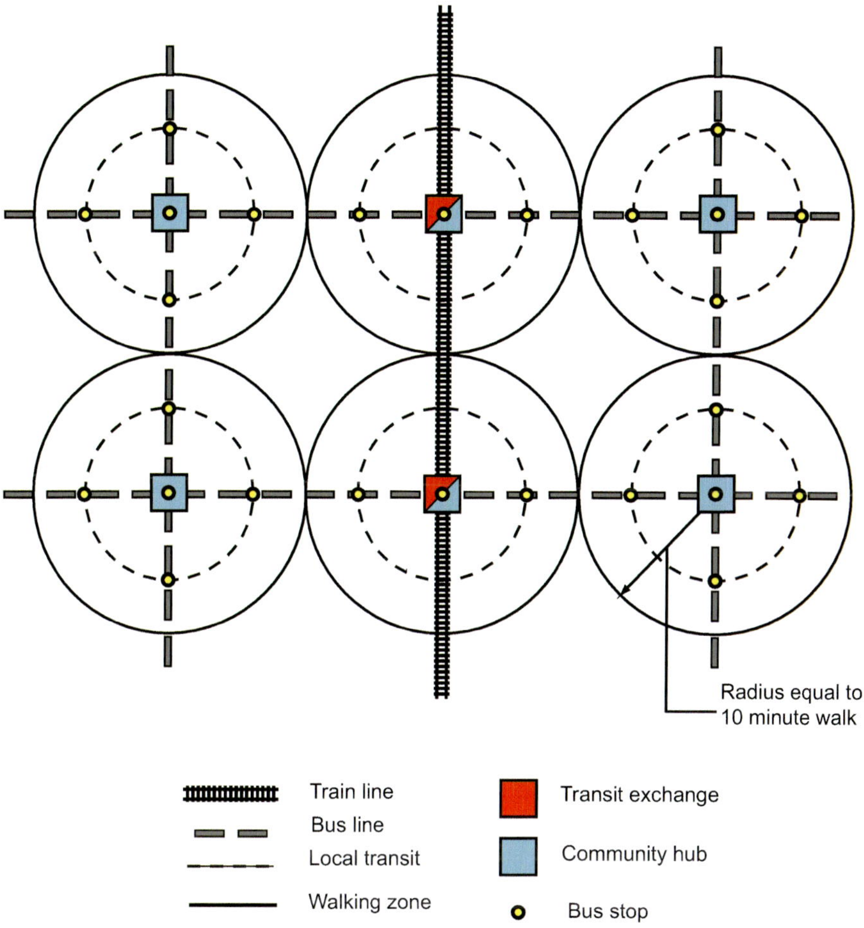

Fig. 15.6 Successful mobility systems include well integrated intermodal means of transit and transit-oriented developments

Optimizing the relationship between land-use and mobility planning is important for successful active transportation (Boname et al. 2009; European Heart Network 2008). Transportation planning needs to integrate land-use policies that affects the built environment where daily transportation choices are made (European Heart Network 2008; Commission of the European Committees 2007). For example, roads can be redesigned to allocate more space for pedestrian walkways and bike paths, narrowing the space for vehicles, therefore encouraging active mobility (this will be explored further in Chap. 18) (European Heart Network 2008).

Development of affordable and sustainable public transport solutions, rather than costly ones such as tram and metro systems, should be prioritized. For example, bus rapid transit (BRT) achieves this through the provision of dedicated lanes, with busways and iconic stations typically aligned to the center of the road, off-board

Fig. 15.7 Accommodating cyclists, having safe bike lanes and pedestrian paths are vital if people are to choose to engage in active mobility

fare collection, and fast and frequent operations (ITDP n.d.). BRT systems offer efficient and affordable access to public transportation while making use of existing infrastructure (Commission of the European Committees 2007), and therefore offer a sustainable alternative to building new tram or metro systems.

Making sure that safe and secure system is provided is another key element to consider when planning for active mobility. The quality of the services provided will influence people's motivation for using it. Bike lanes and pedestrian paths must be safe and reliable, protected from the dangers of vehicles for people to choose to engage in active mobility (Fig. 15.7). Even paving, divided lanes, and good lighting are design factors that will motivate people to opt for active mobility (Commission of the European Committees 2007).

Redesigning neighborhoods is another strategy to promote active transportation. As outlined in Chap. 8, designing residential streets for people rather than vehicles, allow for a balance between traffic and communities to be restored. Vehicle speed should be significantly reduced in residential zones, making them safer for children and seniors, encouraging active mobility (European Heart Network 2008).

Soft measures can also be developed, establishing programs that target travel behaviors rather than physical infrastructure. This can be done by creating alternative workplace and school travel plans or car sharing (European Heart Network 2008). Indeed, initiatives such as traffic games, road safety assessments or educational packages may promote walking and biking as modes of active transportation (Commission of the European Committees 2007). Another soft measure which may be used is road user charging. If drivers are forced to reflect on how much they travel by private vehicle by being charged according to their individual trips, they are likely to be discouraged from using their private vehicles (European Heart Network 2008).

Finally, although planning for green modes of transit and active mobility can be challenging, it does also present many development opportunities as well as economic incentives. Such planning should give priority to active mobility rather than private vehicles. Factors such as land use and zoning should be considered to promote neighborhood sustainability as outlined in Chap. 9. Connecting neighborhoods and services with a reliable active mobility network and public transit network will greatly benefit the community by decreasing the collective carbon footprint and promoting healthy lifestyles.

References

Boname P, Zhu J, Matheson C (2009) Efficient cities: the interrelationship between effective rapid transit systems and the optimal utilization of land use entitlements. https://cwf.ca/wp-content/uploads/2015/11/CWF_GoingForGold5_EfficientCities_MAR2009.pdf. Accessed 18 May 2020

Cervero R (2009) Transport infrastructure and global competitiveness: balancing mobility and livability. Ann Am Acad 626:210–224

Commission of the European Committees (2007) Green paper: towards a new culture for urban mobility. European Union, Brussels

European Heart Network (2008) European Heart Network response to the European Commission on the "Green Paper Towards a new culture for urban mobility". http://www.ehnheart.org/publications-and-papers/responses-to-consultations.html?start=20. Accessed 18 May 2020

Government of Canada (2014) Active transportation. https://www.canada.ca/en/public-health/services/being-active/active-transportation.html. Accessed 22 April 2020

Institute for Transport and Development Policy (ITDP) (n.d.) What is BRT? https://www.itdp.org/library/standards-and-guides/the-bus-rapid-transit-standard/what-is-brt/. Accessed 22 April 2020

Chapter 16
Urban Design for Transit-Oriented Development

Abstract For residents of outlying communities, a daily commute by private car is common. The negative ramifications of such a practice are significant and include pollution, traffic congestion, and on-going infrastructure investment. In an era of greater environmental awareness and high fuel costs, the need to consider alternative transit systems has become urgent. Transit-Oriented Development (TOD) is a strategic urban approach that supports the weaving of various modes of commuting with residential and commercial land uses. In recent decades, TOD has emerged as a leading concept in responding to some of the challenges that cities located on transit corridors face in terms of mobility and housing their citizens. This chapter introduces TOD and illustrates efficient approaches to the urban design of such developments.

Keywords Transit-Oriented Development (TOD) · Affordable housing · Energy-efficient · Cost-effective transportation · Transportation policy · Multimodal methods · Density · Diversity · Housing · Connectivity · Mixed land use · Walkability · Mixed housing types

16.1 The Rational for and History of TOD

As we outlined in Sect. 1.3.1, it is important to consider different methods of public transit when designing affordable, accessible, sustainable, and economically viable alternative to travel by private vehicles. Such methods must be planned and designed in a way which incorporated public transit infrastructure with the existing amenities found in the built environment. Well planned TOD will link and coordinate public transit with various sectors of the city or neighborhood such as residential and commercial areas. TOD encourages economic development by providing transit users easy access to different sectors, promoting job creation and the increased availability of affordable housing. Successful TODs go beyond planning for a transportation system and its physical design, as economic and social elements also determine the proper functioning of such developments (Brinklow 2010).

A. Friedman, *Fundamentals of Sustainable Urban Design*,
https://doi.org/10.1007/978-3-030-60865-1_16

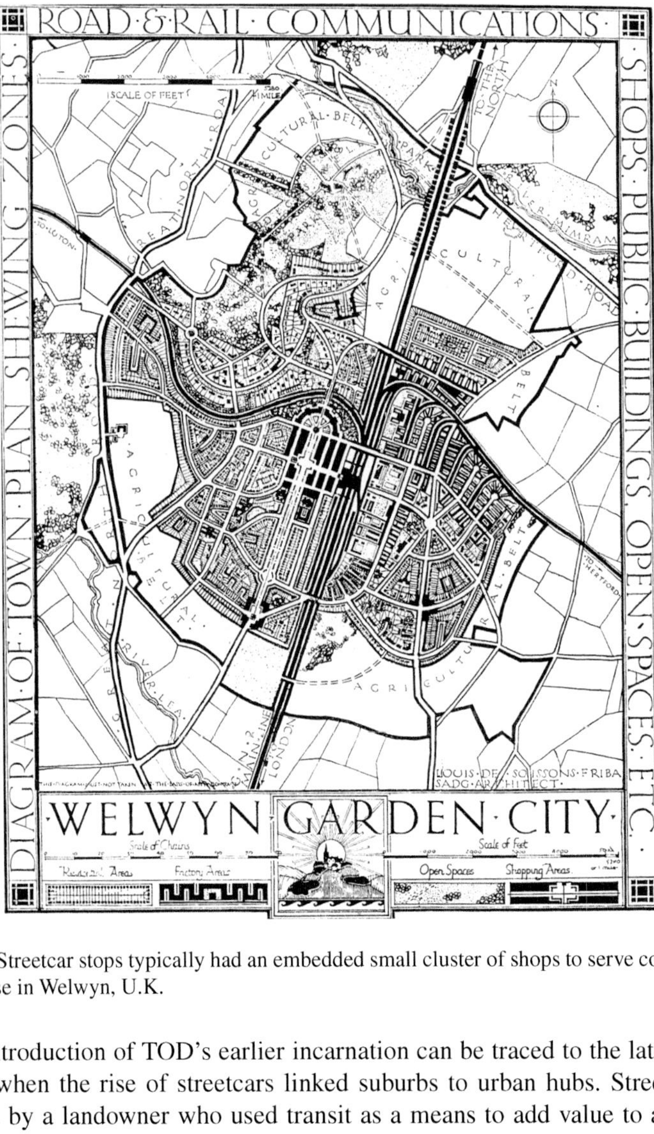

Fig. 16.1 Streetcar stops typically had an embedded small cluster of shops to serve commuters as was the case in Welwyn, U.K.

The introduction of TOD's earlier incarnation can be traced to the late nineteen century, when the rise of streetcars linked suburbs to urban hubs. Streetcar lines were laid by a landowner who used transit as a means to add value to an area by providing a link of commute between jobs in the center and housing at the periphery. Streetcar stops typically had an embedded small cluster of shops to serve commuters, but with the rise of the automobile in the 1930s this type of development was replaced with roads and highways as was the case in Welwyn, UK (Fig. 16.1).

In the mid-1990s, TOD was reintroduced by planner Peter Calthorpe and again in 2004, following the founding of the Center for Transit-Oriented Development and its ensuing projects (Dittmar and Ohland 2004). The advantage is that TOD projects have been found to be more energy-efficient than conventional development practices as illustrated in Fig. 16.2 (Hodges 2010). Commuting costs may be

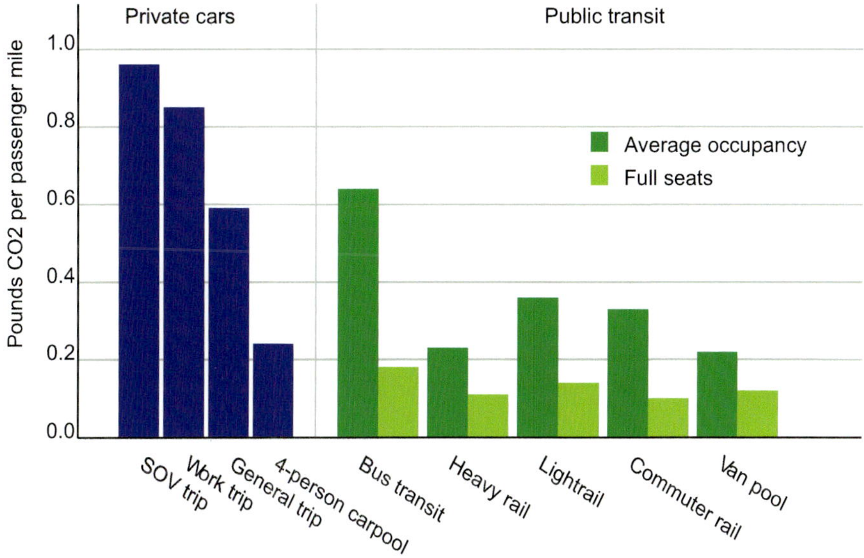

Fig. 16.2 Tailpipe emissions of common transport modes per passenger mile demonstrates the environmental advantage of public transit

reduced by up to 50% if proper access to good transit services is provided, which would especially benefit lower income households.

Calthorpe's approach is unique because he used transportation as a principle design tool. TODs can be described as a planning pattern whereby communities are built along transportation arteries to provide cost-effective transportation, able to link outlaying communities to large centers, which in turn supports the prosperity of such places. Express buses—at-grade, underground, or above-ground rail—or ferries are all common modes of transit which may be found in such developments (Fig. 16.3). The ease of walking and biking, the presence of retail amenities and performance of the housing market in an area determines the potential success of TODs and are important elements to consider when planning. Calthorpe supports the efficient integration between land use and transportation policy, which is the quality distinguishing TODs from other development practices (Renne and Wells 2004).

This strategic urban approach argues for residential and commercial land uses to be planned according to transportation modes (Calthorpe 1993) which results in communities where people can live, work and spend their leisure time. As discussed in Chap. 9, combining mixed-use development and TOD will increase densification and provide easy access to all amenities that might be required. Planners should prioritize multimodal methods of transportation, to reduce the number of trips by private vehicle.

Fig. 16.3 A Bus Rapid Transit stop near Vijfhuizen, the Netherlands

16.2 Urban Design Strategies for TOD

To avoid TODs becoming "transit-adjacent developments" (TADs), certain aspects must be considered (Brinklow 2010). First, ensuring the development is relevant to the surrounding landscape by connecting different modes of transportation between municipalities is of importance. Governments must work together to create comprehensive policies that link land-use policies with transit agencies (Brinklow 2010). Staying true to Calthorpe's guidelines to promote higher densities and mixed land use is also key to successful TODs and should be at the basis of any design.

As many economists argue, a system is only as good as its number of users (Hall 2008), which is why accessibility is fundamental to the success of public transit and active mobility systems. Accessibility is be supported by the three physical principles to create successful TODs, density, diversity, and design. Density impacts the number and length of trips users may need to take to reach their destination. Diversity of development provides users with a variety of services or amenities, preventing the need for unnecessary trips. The design of a TOD should encourage active mobility, specifically walking (Brinklow 2010). Planning for housing, connectivity, mixed land use, and walkability are other elements which should be given as much attention as transportation when coming up with a TOD design scheme. When successful, there will be well-designed community spaces alongside efficient public transit and active mobility systems.

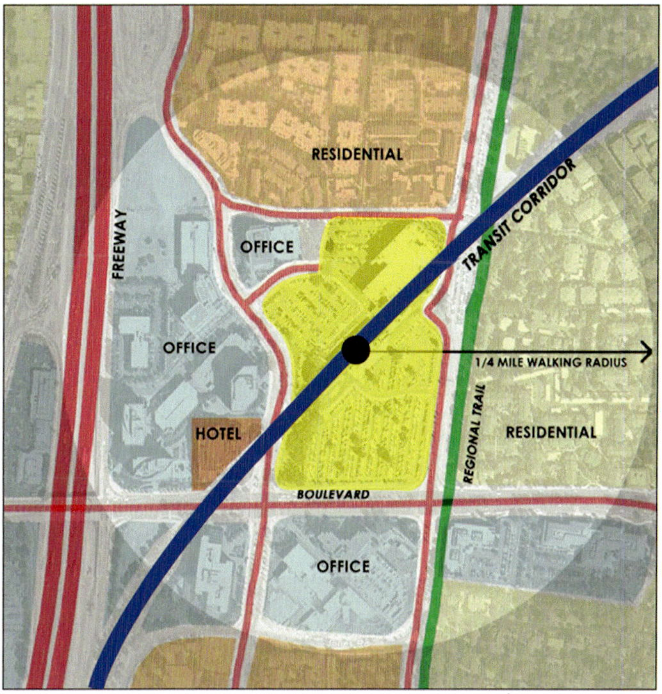

Fig. 16.4 The transit-oriented Contra Costa Centre Village is located 48 km (30 miles) North-East of San Francisco, California, US and serves as a connection point between three cities

The Contra Costa Centre Village, USA, illustrates how transit systems can be used to spawn sustainable and diverse communities. Located 48 km (30 miles) north-east of San Francisco, it serves as a connection point between three cities: Walnut Creek, Pleasant Hill, and Concord. Prior to the TOD effort, this area was semirural, and comprised mainly of single-family, low-density housing.

TOD was introduced to Contra Costa by way of the Bay Area Rapid Transit (BART) system—a 167 km (104 mile) light-rail track that serves 44 stations. Contra Costa station is currently one of the area's most important commuter-hubs; it consistently has the greatest number of morning passengers and serves 350,000 people annually. In the 1970s, developers recognized the importance of this station and attempted to create a transit village on the surrounding land. However, due to various complications, the project was only recently realized (Fig. 16.4).

Planning TODs near existing residential developments as well as the urban design of these places is an important aspect to consider (Fig. 16.5). Providing housing near TOD is naturally a key component to the development's success as it ensures a substantial ridership. However, the nature of residential developments is also worth considering. For instance, planning for mixed housing types plays a significant role to ensure that TODs are resilient when there are fluctuations in the housing demand, as providing multiple housing options allows more individuals to

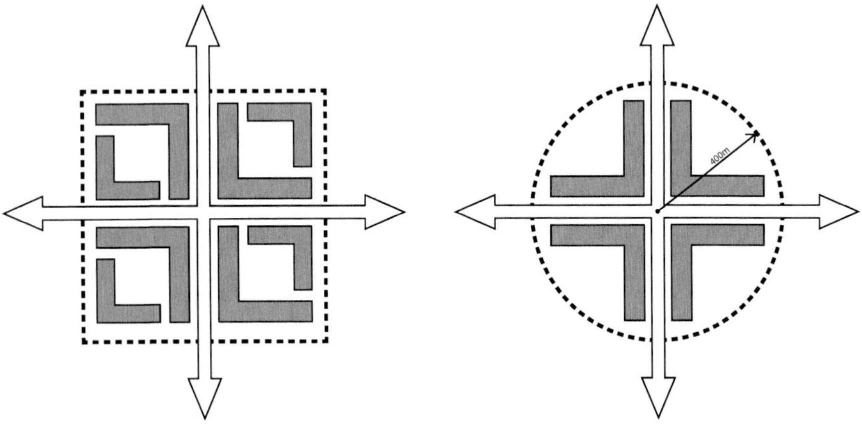

Fig. 16.5 Possible relations between and transit lines and buildings in transit-oriented developments (TOD)

access housing. Furthermore, housing densities do matter; Calthorpe established certain guidelines for successful housing densities surrounding TODs. Housing near a transit hub should be moderate- to high-density and be located in a mixed-use development. On average, there should be 18 dwellings per acre (7 dwellings per hectare), with areas closer to the transit hub having a higher concentration of 26 dwellings per acre (10.5 dwellings per hectare); areas at the perimeter of the development having around 10 dwellings per acre (4 dwellings per hectare) and including small-lot family homes, which further contributes to the presence of mixed housing types.

Planning for proper linkage between existing features of the built environment and TODs is crucial for their success. Existing sites may be good locations for TODs to be built through redevelopment, as underdeveloped sites or infill sites offer great opportunities for developers to transform car-centric places into areas where public transit and active mobility are offered. New urban areas are also sites where TODs may be built and are often located nearby existing communities which will make new transit development available for previous and new residents, increasing the volume of users.

Three transit corridor types are considered when designing TODs: destination connector, commuter, and district circulator. Destination connectors accommodate ridership in both directions during the day, as they serve key employment, public, and residential destinations. Commuter corridors link major hubs found within a larger region, and district circulators run public transit loops within districts. Planning TODs at the corridor level is important not only for their contribution to the region, but for the towns that find themselves along the corridors as well. While destination connectors are efficient at accommodating ridership throughout the day,

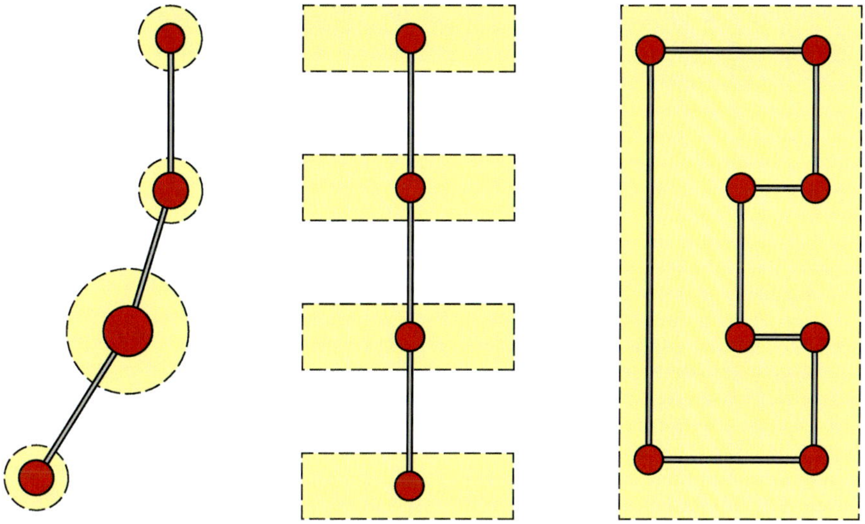

Fig. 16.6 The three transit-oriented developments (TOD) corridors are: destination connectors (left), commuter (middle) and district circulator (left)

commuter corridors connect important hubs within a wider region and district circulators concern themselves with public transit within districts (Fig. 16.6).

Mixed zoning is another key feature which should be prioritized in every TOD design scheme. Commerce, employment, and public spaces should be located at the core so that growth may occur around transit stops. Furthermore, pairing high-speed innovations with TOD designs such as BRT or high-speed trains, scheduled to run frequently during rush hours, will result in positively sustainable impacts in peripheral areas as individuals choose to stay in less urbanized areas (Deng and Nelson 2011).

Planning for walkability is another important element to consider when planning TODs as neighborhoods which promote walking almost naturally result in ones where transit use is more heavily relied upon (Brinklow 2010). Streets should be appealing and safe to pedestrians so that they are encouraged to walk. Furthermore, dwellings, basic amenities, employment, and entertainment hubs should be within a walkable distance from transit stops so that walking may present itself as an accessible mode of transportation. This will not only promote the use of active mobility but will also make TODs more resilient.

Building advanced transit systems such as TODs is costly but should be considered a cost-effective investment. This is why it is important to make sure TODs are planned in a way that will maximize use, as advantages may follow from such developments including job creation and more affordable housing. Therefore, there is motivation and incentive to invest in such projects, but these must be organized for them to present as many benefits as possible. Planning for TODs is more than designing a transportation system, it is a design approach that considers transportation as a design tool. TOD involves much coordination and a broad perspective considering the opportunities as well as the challenges that are found in the landscape, in order to fit new transit developments in accordance with what is already visible or not yet visible in the surrounding environment.

References

Brinklow A (2010) Transit-oriented development: a policy implementation strategy. Master of Urban Planning thesis. McGill University, Montreal

Calthorpe P (1993) The next American Metropolis. Princeton Architectural Press, New York

Deng T, Nelson JD (2011) Recent developments in bus rapid transit: a review of the literature. Transp Rev 31(1):69–96

Dittmar H, Ohland G (eds) (2004) The new transit town: best practices. In: The new transit town. Island Press, Washington, DC

Hall P (2008) Urban renaissance, urban villages, smart growth: find the differences. In: Haas T (ed) New urbanism and beyond: designing cities for the future. Rizzoli, New York

Hodges T (2010) Public transportations role in responding to climate change. Diane Publishing, Darby

Renne JL, Wells JS (2004) Emerging European-style planning in the USA: transit-oriented development. World Transport Pol Pract 10(2):12–24

Chapter 17
Alternative Standards for Streets, Paths, and Pavements

Abstract The rapid residential development that followed World War II saw a gradual increase of street width. Parallel to their expansion was an increase in the number of utilities that ran alongside or beneath them. The oversized street with buried utilities now accounts for fully 25% of a home's cost; maintenance, road resurfacing, cleaning, and snow removal have become a significant part of the municipal tax bill. This chapter investigates alternative roads, paths dimensions and ground cover for cost and land saving, as well as overall contribution to sustainability.

Keywords Transit-Oriented Development (TOD) · Affordable housing · Energy-efficient · Cost-effective transportation · Transportation policy · Multimodal methods · Density · Diversity · Housing · Connectivity · Mixed land use · Walkability · Mixed housing types

17.1 Rethinking Current Standards

Designing streets to accommodate automobile traffic has been a common practice in urban design since the mid-twentieth century (Fig. 17.1). Although it is difficult to undo a practice which has been used for a long time, it is important to introduce alternative standards that shift the focus from designing streets for cars to designing streets for active and public transit.

Introducing alternative standards is the first step toward striving to accomplish such a shift and may be done by designing streets at the human scale. Attempts to design and document the benefits of narrow streets have already been undertaken in the field of urban design. These attempts started by establishing relationship between street function, street width, and how it fits in the larger local street network. Designing narrow streets with this relationship in mind is therefore key, as they must be made part of a network which had previously overlooked their economic, environmental, and social benefits.

Generally, it is important to note that in most cases, large road widths are inefficient and unsafe. For instance, neighborhoods commonly allocate around 9 m

A. Friedman, *Fundamentals of Sustainable Urban Design*,
https://doi.org/10.1007/978-3-030-60865-1_17

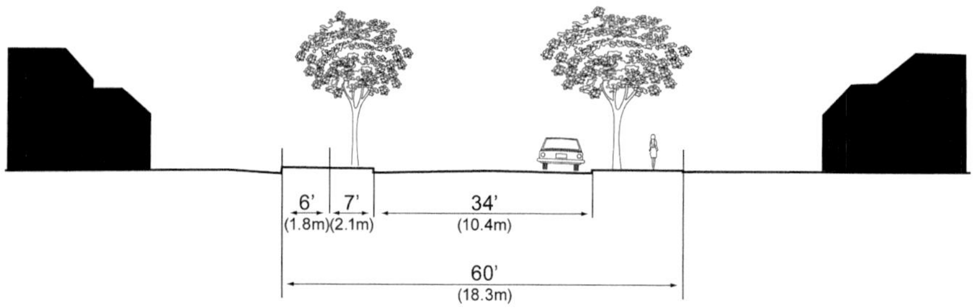

Fig. 17.1 Designing streets with wide right of way to accommodate vehicular traffic has been a common practice in urban planning since the mid-twentieth century

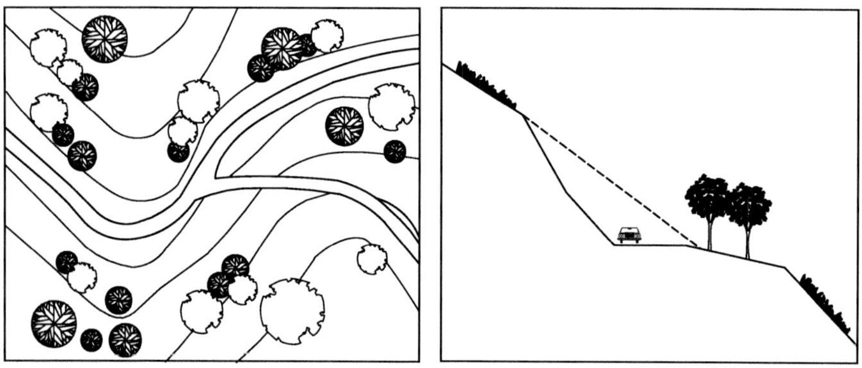

Fig. 17.2 The existing natural environment needs to be given greater consideration when planning streets like having roads along contour line

(29.5 ft) for a road width and 3 m (10 ft) to sidewalks. However, two parallel vehicles need only a minimum of 30 cm (1 ft) between them if they are driving at 32 km per hour (20 miles per hour) (Southworth 1997). Not only do wider streets encourage vehicles to travel faster, but they also take away from the presence of other amenities, such as green spaces, by monopolizing significant amounts of land. Narrower streets have been found to be more effective, as they use less valuable land and lower construction costs. The existing natural environment needs also to be given greater consideration when planning streets like having roads along contour line (Fig. 17.2).

Narrow streets promote social objectives since many health benefits may be associated with such designs, like the use active transportation. Narrower streets diminish the speed and volume of cars which makes them safer to walk and cycle on, and access to public transit options (see Chap. 18 for further detail). Furthermore, if streets are planned with multiple uses in mind and not solely to provide a path for private motorized vehicles to flow through, they have the potential to promote other social aspects of cities such as feelings of belonging and increased neighborliness (Fig. 17.3).

Fig. 17.3 Properly scaled well-designed streets can serve dual purpose and promote feelings of belonging and, increased neighborliness as is the case in Singapore (left) and Bruges, Belgium (right)

17.2 Designing Narrow Streets, Lanes, and Paths

Designing streets with a social objective often results in narrower streets. This perspective is encouraged by Michael Southworth (1997) who argues for the integration of roads into living environments because social activities and mobility patterns occur on pavement. Planners should design residential roads for cars and pedestrians and include features such as street furniture and plants. This practice naturally results in the beautification of the street and the sidewalk, slowed traffic, and the creation of much narrower streets (Fig. 17.4).

Similarly, Dan Burden (2001) uses the term "walkable scale" as a guiding principal to design neighborhoods with pedestrians in mind. The focus on narrow streets is indeed an important element of this design concept. For instance, the width of streets should be reduced alongside the reduction of car lanes to make room for bike and pedestrian paths (Fig. 17.5). Neighborhood design should include short blocks, slow down cars, and make walking safer and more convenient.

17.3 Alternative Ground Cover

Although the most widely used material to surface streets is asphalt, it is not the most sustainable building material. However, there are more sustainable options such as permeable pavement technologies, interlocking pavement, brick, and vegetation.

There are two types of pavements: rigid and flexible pavements. Rigid pavements are thicker and are usually built to last for 30 to 40 years. The last layer is one of concrete slab which is able to absorb the stress of traffic and therefore does not require the need for many layers beneath it. Flexible pavements are composed of multiple layers needed to offer enough support for traffic loads and are built to last

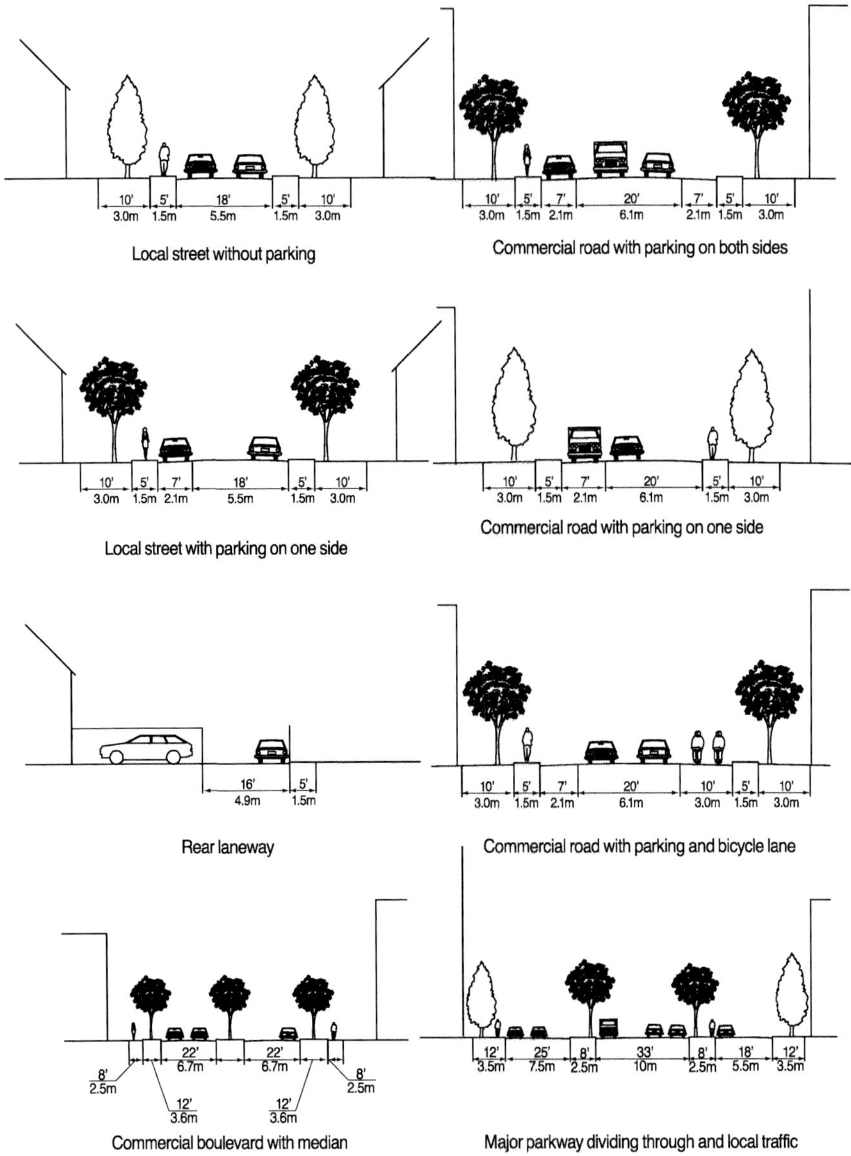

Fig. 17.4 Alternative dimensions have been suggested by various municipalities for roads to make them more sustainable and cost-effective

between 15 to 20 years, as their structure is less resistant to deterioration. The surface layer typically consists of one or two bituminous or Hot Mix Asphalt (HMA) (Chandra n.d.). The use of either one is determined by the existing condition of the terrain, weather and traffic conditions of the area. Although concrete was the preferred choice before, asphalt the main material used today to pave roads and streets,

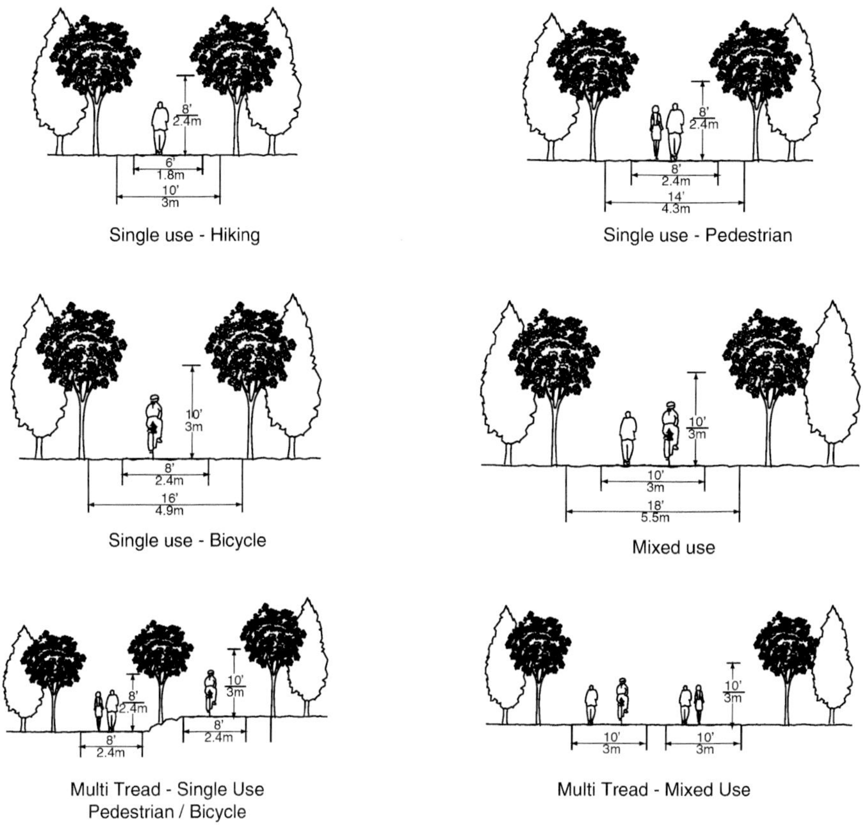

Fig. 17.5 Streets' right of way should be reduced to make room for bicycles and pedestrian paths

as it offers more flexibility and is less energy intensive when making alterations or repairs (ANS Construction n.d.).

However, maintenance is costly and has negative environmental impacts which is why pavement technology may offer a good alternative to asphalt. For instance, two-polymer modified Superpave mixtures, one with optimum amount of binder and another with rich binder content have been shown to offer more long-lasting pavement quality than conventional asphalt pavements (Priyanka et al. 2019).

Another alternative is Interlocking Concrete Block Pavement (ICBP) which consist of small precast concrete blocks used to cover street surfaces, which make street covers easier to maintain as these are easy to remove and replace (Fig. 17.6) (Sharma n.d.). Furthermore, interlocking does not necessarily have to be done using concrete blocks, as blocks made of brick rather than asphalt are also resilient such as the paving blocks from Unipaver. Although they are susceptible to deterioration, interlocking pavements are easier to maintain than a large slab of concrete or asphalt for instance, as the bricks can simply be replaced by new ones. Paving blocks also offer a variety of pavement design as they include different styles and shapes (Samanvi

Fig. 17.6 The use of permeable Interlocking Concrete Block Pavement (ICBP) which consist of small precast concrete units makes streets surface easier to maintain and replace

Technologies 2020). This may contribute to making streets more attractive and encourage the use of active modes of transportation.

Finally, vegetation is another element which should be included in street cover design. Vegetation can contribute to low impact design, which manages storm water at the source, and may also alleviate long-term maintenance costs (Fig. 17.7). For instance, including vegetation between the street and the curb would help reduce the accumulation of water at pedestrian ramps and could act as both buffers against traffic as well as attractive features to encourage people to walk and use public transit. Furthermore, streets would have to be narrower by default as the addition of trees or other plants would require space (NACTO 2012).

A combination of designs for narrower streets and the use of alternative ground covers would make streets, paths, and pavements more sustainable and further encourage sustainable behavior by promoting the use of active transportation. Designing them considering the needs of pedestrians rather than private motorized vehicles, which have been the main preoccupation in street and road designs in the past decades, is the first step toward making them more sustainable in all economic, environmental, and social aspects.

Fig. 17.7 Permeable surfaces with vegetation let rainwater penetrate and reach soil rather than end up in a storm sewer

References

ANS Construction LLC (n.d.) Why should you choose asphalt instead of concrete? https://ansconstructionllc.com/choose-asphalt-instead-concrete/. Accessed 11 Feb 2020

Burden D (2001) Building communities with transportation. https://www.lgc.org/wordpress/docs/freepub/community_design/presentations/dan_burden_2001_pres.pdf. Accessed 17 May 2020

Chandra S (n.d.) Flexible pavement versus rigid pavement. https://www.nbmcw.com/tech-articles/roads-and-pavements/36977-flexible-pavement-versus-rigid-pavement.html. Accessed 11 Feb 2020

National Association of City Transportation Officials (NACTO) (2012) Urban street design guide. https://nacto.org/publication/urban-street-design-guide/. Accessed 11 Feb 2020

Priyanka BA, Sarang G, Ravi Shankar AU (2019) Evaluation of Superpave mixtures for perpetual asphalt pavements. Road Mater Pave Des 20(8):1952–1965

Samanvi Technologies (2020) Unipaver. http://www.unipaver.in/. Accessed 11 Feb 2020

Sharma S (n.d.) Interlocking Concrete Paver Blocks. https://www.nbmcw.com/tech-articles/concrete/4993-interlocking-concrete-paver-blocks.html. Accessed 11 Feb 2020

Southworth M (1997) Walkable suburbs? An evaluation of neotraditional communities at the urban edge. J Am Plann Assoc 63(1):28–44

Chapter 18
Urban Design for Safe Walking and Biking

Abstract While the use of automobiles as a means of transport is often associated with negative environmental, economic, and health outcomes, walking and cycling are modes of active transportation which offer an array of benefits. Solutions aimed toward improving features of the built environment through simple measures such as paths which are well-indicated and proper signage will significantly contribute to the safety of pedestrians and cyclists and can generate a cultural shift from car use to walking and cycling (Winters et al., J Urban Health 87:969–993, 2010).

Keywords Pedestrians · Cyclists · Livable places · Traffic congestion signals · Hierarchical system · Divided bike paths · Woonerf · Active mobility · Residential density urban landscapes · Sidewalks · Pathways · Crosswalks · Social interaction · Freeway service roads

18.1 Measures of Slowing and Reducing Vehicular Traffic

There are many benefits associated with designing urban landscapes to promote safe active mobility. For example, many cities pedestrianized their centers or converted them to places with shared traffic (Fig. 18.1). Those with high walking and cycling modals have reported positive economic impacts, an improved quality of life for residents, less roadway congestion, and better health (European Cyclists' Federation 2009). It also makes neighborhoods safer, more social, and more livable places. To design for safe walking and biking in urban environments, adjustments must first be made so that the presence of private vehicles is reduced and is no longer prioritized as the principle mode of transportation. After doing so, planners should consider the needs of pedestrians and cyclists, as the safety and ease of using active mobility networks must be enhanced. Walking and biking can be made more accessible and appealing by first discouraging the use of cars and then by encouraging the design of features in the built environment catered to the needs and comfort of pedestrians and cyclists.

In most cities, there is a need for better coordination between vehicular traffic, pedestrians, and cyclists (Jacobsen et al. 2009). Traffic congestion signals a need for

A. Friedman, *Fundamentals of Sustainable Urban Design*,
https://doi.org/10.1007/978-3-030-60865-1_18

Fig. 18.1 Many cities such as Beijing, China (top left), Barcelona, Spain (top right), Lodz, Poland (bottom left), and Cologne, Germany (bottom right) made part of their centers exclusively pedestrian or introduced shared traffic to make them more walkable

cities to stop building large roads, and to rather plan environments at the human scale (Low 2003). It is for this reason that discouraging the use of vehicles is important. This may be done using certain strategies to reduce traffic intensity and speed which have been shown to influence the perceived safety of other modes of active transportation. Slowing down vehicular traffic may be achieved by either installing physical buffers or by implementing regulations through zoning and signaling (Fig. 18.2).

In order to reduce car use and the traffic, a hierarchical system must be set up whereby pedestrians and cyclists find themselves prioritized in transport policies, followed by public transportation, and finally by vehicles (European Cyclists' Federation 2009). Reallocating space is a clear way to implement that hierarchy; transportation modes that use less space should be prioritized in the urban landscape that will naturally result in less vehicular traffic. For instance, widening walkways and developing divided bike paths effectively reclaims street space that was previously used by vehicles. This is in accordance with Nicholas Low's idea of the "active city" whereby car dominance in central areas is reduced as urban spaces are planned in a way to prioritize walking, cycling, and the use of public transportation (Low 2003).

As noted above, designing streets which consider pedestrians and cyclists rather than just private motorized vehicles can be successful in encouraging active modes

Fig. 18.2 Slowing down vehicular traffic may be achieved by either installing physical buffers or by implementing regulations through zoning and signaling

of transportation such as walking and cycling. Streets such as the woonerf (meaning street for living in Dutch) in the Netherlands is narrow and include curves which slow down car traffic, making walking and cycling safer. The woonerf street will be further explored in Chap. 19.

Driving private vehicles can be discouraged by lowering speed limits and automobile taxation. Implementing 15 to 30 km per hour zones (9 to 18 miles per hour) and congestion charging (e.g., a daily charge for driving a vehicle within a defined zone during set hours) are successful tools to slow down traffic and discourage the use of vehicles (Jacobsen et al. 2009). Road user charging has been successful in some cities including London, Stockholm, and Milan which experienced a reduction in vehicle use and an increase in the rates of active mobility in zones where charging was implemented (European Cyclists' Federation 2009). Furthermore, strategically zoned parking has also shown to contribute to reducing vehicular traffic.

Physical interventions in the built environment of dense areas such as road bumps or highly textured driving surfaces are also successful methods to slow vehicular traffic (Fig. 18.3). The bumps can be placed at the entrance of a street to indicate an increase in residential density and act as a warning sign for drivers to slow down. Stamped concrete or cobblestone segments can also be effective, emphasizing the urgency to slow down near gateways and entrances. Raising the level of the road at intersections is another means of decreasing heavy vehicular traffic on certain streets, as vehicles would necessarily have to slow down when approaching an intersection.

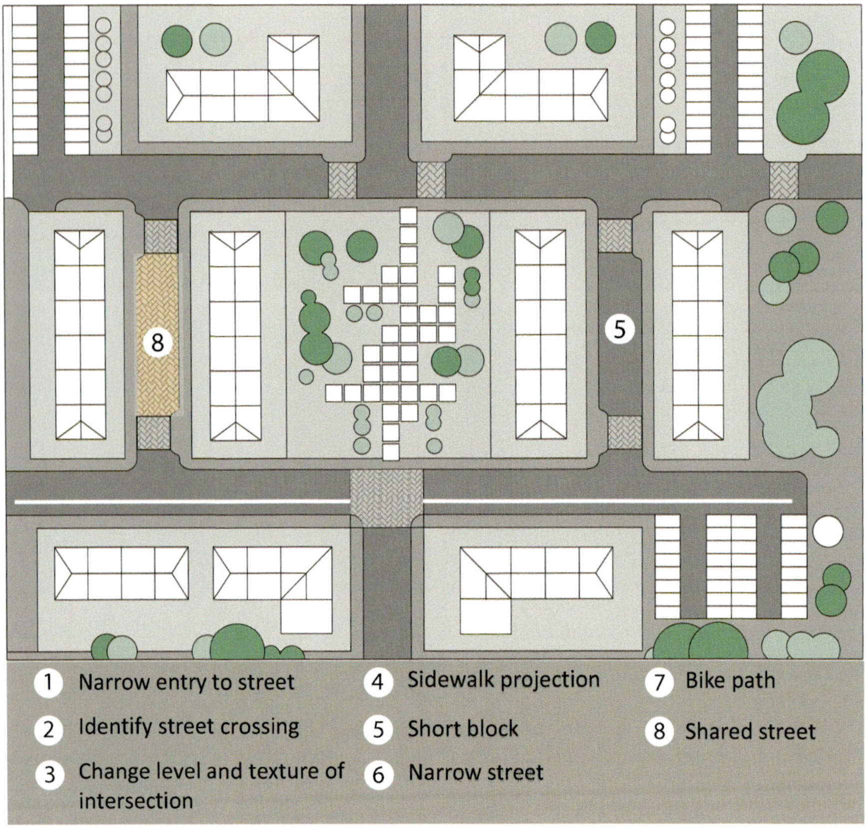

Fig. 18.3 Design of safer streets for pedestrians and cyclists include measures such as change level altering the texture of intersections and creating sidewalk projections

18.2 Safer and Appealing Walking and Cycling

Promoting environments to encourage walking and cycling can be achieved by including certain features in the built environment or planning and managing urban landscapes in a certain way. By doing so, other modes of active mobility will be encouraged and made more accessible as they are integrated into road networks.

Sidewalks, pathways, and crosswalks— the section of urban landscapes allocated to pedestrians— must be designed in a way to make walking safe and convenient; integrating pedestrian movement into road and street networks is key to doing so. Footpath designs should consider certain factors for them to be successful and cater to pedestrians' needs. There are pedestrian systems categorized by three types of footpaths designed and built in addition to sidewalks. Such systems should be considered so that footpaths may be strategically positioned in new projects based on their intended use which also determines their width. Walkways link dwellings to daily utilitarian functions such as parking. There are then paths found between buildings, followed by paths in networks connecting homes with community amenities

Fig. 18.4 Foot paths can be located between buildings, part of networks connecting homes with community amenities such as parks and schools, as well as with commercial areas as shown in Montreal, Canada (top left), Victoria, Canada (top right), Kfar Sava, Israel (bottom left), and Berlin, Germany (bottom right)

such as parks and schools, as well as with commercial areas (Fig. 18.4). All footpaths must be designed with pedestrians' needs in mind and provide comfortable environments. For instance, paths should not be located in isolated or unsheltered areas, as these could create social and health safety issues, deterring people from using them.

Pedestrians of all ages and conditions should be considered and paths should be designed to accommodate as many types of users as possible (Fig. 18.5). Park benches should be readily available along paths and curb cuts should be present to make footpaths accessible to wheelchairs or strollers. Adequate lighting is important to ensure footpaths can be used day and night. Providing continuous and uninterrupted sidewalks also promotes safety. Regular maintenance such as snow removal and repairs should be done regularly to ensure paths are always safe and accessible. Street furniture such as benches are also key elements to foster walking habits by making walking a more enjoyable experience and promoting social interaction.

To make walking more convenient, pedestrian networks should be designed with other transportation networks in mind so that there is an easy transition from one mode of transport to another. Directness of routes is a key aspect which greatly influences walking habits, as the presence of fewer obstacles to reach a certain destination will encourage people to walk more. Paths should be built along the quickest route. If it is becomes known that residents prefer alternate paths, those should be included in the network. Footbridges can be used to reduce the number of obstacles on a given route.

Fig. 18.5 Pedestrians of all ages and conditions should be considered in designing paths like this one near Montreal, Quebec, Canada

Fig. 18.6 Special signs can be installed near intersections alerting motorists that elderly population resides in the area like this one in Melbourne, Australia

Intersections between footpaths and roads should be given extra attention in the planning process as they can be hazardous. They can be made safer by implementing narrower street so that intersections are smaller and crossing distances are shorter. In addition, signs can be installed alerting motorists to be aware of pedestrians, especially children and seniors (Fig. 18.6). In the case of wider intersections with divided lanes, there should be pedestrian refuges and waiting areas where pedestrians can stay until it is safe to cross. These shelters should be placed at the median between the lanes so that slower walkers have a place to safely wait if they do not have enough time to cross entire lanes. They should be minimum 1.2 m (3.6 ft) wide and extend beyond the crosswalk so that pedestrians are protected from left-turning traffic. The addition of wide sidewalks separated from traffic lanes by planted strips may protect pedestrians from vehicular traffic while also softening the effect of hard asphalt surfaces. In addition to the physical adaptations, the time allocated for pedestrians to cross a road should be extended.

There are many strategies which can be implemented to make urban landscapes safer and more relevant to cyclists' needs, increasing cyclist safety and promoting active mobility. Much like well-designed pedestrian networks, well-built bike paths should link different key urban nodes to one another; commonly used services and amenities should be part of the system.

Making sure there is enough accommodation of cyclists' needs is a key element to consider, as well as ensuring that such infrastructure is integrated with public transportation (Fig. 18.7). This may be done by narrowing vehicles' traffic lanes or converting some roads to one-way streets so that the remaining space may be used to add bike lanes. Much like the design of road networks, bike networks should also be hierarchical in nature. There should be bicycle freeways which may be established on old railroads, utility easements, freeway service roads, river courses, and secondary or very lightly traveled roads.

Bike routes along arterial roads can be the most efficient way to provide easy access to key areas and encourage active mobility. However, to make bike paths

Fig. 18.7 Introducing amenities for cyclists' needs like bikes on buses, storage facilities, traffic lights and a stand with maintenance tools need to be key consideration when planning an effective bicycle network

safe, it is important that sections along arterial roads are physically separated from vehicle traffic. Bike paths should be at least 1.5 m (5 ft) wide and should be clearly marked with signs and paint, clearly separating them from traffic (NACTO 2014). Unlike pedestrian paths, bike paths should be designed to cover longer distances and should be planned to connect users to amenities found in the region such as sport centers or schools. Bike rental stations and racks should also be placed near such amenities for convenience and to ensure ease of use (Fig. 18.8). Encouraging

Fig. 18.8 Bike rental stations need to be placed among other spots near amenity hubs or transit stops for convenience

cycling through implementing affordable public bike systems is another way to encourage people to opt for active mobility rather than using vehicles (see Chap. 21).

There are many strategies which may be implemented to encourage walking and biking in urban environments that aim to both discourage the use of private vehicles and encourage the use of active modes of transportation by making them safer and more accessible. Although these strategies range from restructuring parts of the built environment to implementing new programs and regulations, they should all be considered when planning urban landscapes for safe walking and biking.

References

European's Cyclist Federation (2009) Future cities are cycling cities! https://ecf.com/sites/ecf.com/files/Future-cities-are-cycling-cities.pdf. Accessed 18 May 2020

Jacobsen PL, Racioppi F, Rutter H (2009) Who owns the roads? How motorized traffic discourages walking and bicycling. Inj Prev 15(6):369–373

Low N (2003) The active City. Urban Policy Res 21(1):5–7

National Association of City Transportation Officials (NACTO) (2014) Conventional bike lanes. https://nacto.org/publication/urban-bikeway-design-guide/bike-lanes/conventional-bike-lanes/#. Accessed 23 Apr 2020

Winters M, Brauer M, Setton EM, Teschke K (2010) Built environment influences on healthy transportation choices: bicycling versus driving. J Urban Health 87(6):969–993

Chapter 19
Car-Free Environments and Shared Streets

Abstract Car-free places are a sustainable solution to the increasing financial and environmental costs of fuel related to the heavy reliance on private vehicles. Improved quality of life and public health are important benefits which result from walkable, car-free designs. This chapter discusses how communities can reduce car traffic and relying instead on walking, biking, and the use of public transit. The methods explored outline how to integrate streets, pedestrian paths, parking spaces, and public transit to create a desirable, walkable environment.

Keywords Car-free · Integrate streets · Pedestrian paths · Parking spaces · Public transit · Walkable environment · Public transit networks · Grouped parking · Shared street · Sense of place · Integrated curved

19.1 Key Features of Car-Free Environments

Certain elements can be considered when planning car-free environments, so that modes of active transportation are easy to embrace and made accessible to a wide range of people. Placing key amenities or services closer to where people reside, designing well-integrated public transit networks within the neighborhood, limiting parking, and adopting measures to foster active mobility are ways to decrease car dependency.

The siting of key amenities and services must be done to ensure such destinations are easy to reach by foot or a bicycle. There are different strategies which planners may adopt to bring these destinations closer to where people live. Clustering key neighborhood amenities and locating them strategically is a first step. For instance, placing schools, libraries, or sport facilities at the center of neighborhoods makes the centers easy to reach and may foster a culture of bike riding and walking among children (Fig. 19.1). Furthermore, planning for mixed-use whereby commercial and residential uses are combined or found in close proximity, is another way to help make key destinations faster and easier to reach.

Integrating public transit within the neighborhood and making sure it is relevant to local needs is important for it to be accessible to walkers and bikers. Designing

A. Friedman, *Fundamentals of Sustainable Urban Design*,
https://doi.org/10.1007/978-3-030-60865-1_19

Fig. 19.1 A school over a public library in Ørestad near Copenhagen, Denmark

public transit networks must be organized in a hierarchical way based on demand. For instance, the network can be planned to allow residents to reach buses or trains connecting them to larger urban centers. Buses need to be scheduled to run according to local demand for the transit system to be efficient. Providing more buses during rush hour or ensuring there are buses adapted for seniors or people with disabilities are important factors for a bus networks to be accessible and valuable to as many residents as possible (Fig. 19.2).

Limiting parking through both design and policy measures is a good strategy to discourage the use of private motorized vehicles in neighborhoods. Often, parking spaces dominate urban and suburban areas. While residential parking is typically found at the rear or front of homes, it can serve more than one function and be integrated with the surrounding area. An alternative is to design off-street parking with permeable paving material. Grouped parking is another alternative to single space residential parking whereby a space is designed to accommodate about 20 cars (Fig. 19.3). This is meant to make for maximum flexibility in the arrangement of space and minimize construction and maintenance costs. Furthermore, shared parking lots can become dynamic social spaces. Parking lots can be designed to improve the social environment of communities if visual features are considered. The addition of trees and shrubs can make for a pleasant atmosphere for people to stop and socialize.

As was discussed in Chap. 18, active and public transportation need to be well designed for the success of car-free communities. Biking has become a common way for people to get around, and there has been a large increase in initiatives and

Fig. 19.2 A transit system needs to be inclusive with ramps and adapted buses for riders with reduced mobility like the one shown in Middlebury Vermont, US

funding for policies that encourage biking by both central governments and municipalities (Pucher and Buehler 2008). In addition, walking has been found to greatly benefit public health, which is why planning for walkable neighborhoods is important to consider. Yet, there is no blueprint as to how car-free communities should be designed, as this varies according to the context. There are however certain strategies which may be implemented to plan for successful walkable neighborhoods. These include traffic calming measures, zones with 15 to 30 km per hour (9 to 18 miles per hour) speed limits, congestion charging whereby drivers are charged a daily charge for driving in certain zones and placing bicycle lanes on main streets (Jacobsen et al. 2009). Furthermore, walkers and bikers should be given priority as vulnerable road users to ensure their safety.

19.2 The Dutch Woonerf Experience

The "woonerf" (Dutch for "streets for the living") is a successful street design which allows shared use by pedestrians, cyclists, public transit, and private cars. This street design was first implemented in the Netherlands but is found today in other countries described under various terms such as the "home zone" in Britain or the "shared street" in the United States (Collarte 2012). What distinguishes the woonerf however, is that it is first and foremost an attempt to transform streets into livable spaces and creating a sense of place (Fig. 19.4).

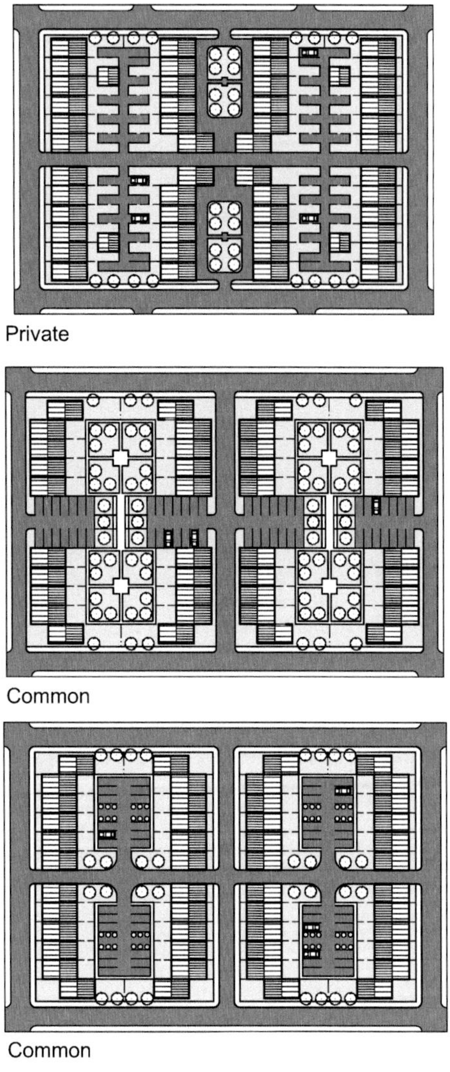

Fig. 19.3 Grouped parking near higher density dwellings is designed to accommodate about 20 cars which minimize construction and maintenance costs

The woonerf originated in the late 1960s when residents opposed heavy car traffic coupled with the need to save on construction costs. Asphalt cover streets and sidewalks were replaced with integrated curved and winding brick streets and limited speed (Collarte 2012).

A notable feature of the woonerf is that there is no difference between sidewalks and car lanes. This makes it so pedestrians are safe to walk on these streets, as cars must necessarily drive at slower speeds. Furthermore, this street design includes narrow streets and soft curves which also contribute to slowing down vehicular traffic. A width of 6 m (20 ft) has been shown to foster safer environments for pedestrians. Slowing and reducing vehicular traffic not only makes pedestrians feel safer, but also allows streets to be used as public spaces (Collarte 2012). The woonerf design includes more street furniture such as trees and benches as well as areas

Fig. 19.4 What distinguishes the woonerf concept, also known as Street for Living, is that in addition to safety and cost saving it is also an attempt to transform streets into livable spaces with a sense of place as shown on the left in Vienna, Austria (left) and Leiden, the Netherlands (right)

where people can gather and interact. Streets therefore become areas which promote livability and where different activities can take place (Collarte 2012).

The four principles of a woonerf street adapted from Eran Ben-Joseph's 1995 article *Changing the Residential Street Scene* are to ensure the street has visible entrances, physical barriers, shared and paved space, as well as landscaping and street furniture (Steinberg 2015). There are clear signs at the entrances of the woonerf so that pedestrians and car users alike are aware of the presence of the streets. All street users are included in the design, as the paved space is intended to be shared and used by everyone. Curves and other physical barriers are important to include to slow down vehicular traffic.

19.3 Learning from Slateford Green

Slateford Green was designed by Hackland and Dore Architects as a car-free community in the Gorgie district of Edinburgh, UK, 3 km (1.9 miles) from the city center with 120 units. A two- to four-storey housing line surrounds gardens in the center of the community, which is itself situated in a natural landscape. Sustainability and efficiency were the first principles of its creation (Fig. 19.5).

The developer created a place that illustrates the benefits of car-free living. Today, six more communities around Edinburgh have been planned that promote a car-free lifestyle. The design also stated that the community must be barrier-free and low maintenance. Thirty-nine of the housing units are shared ownership, 12 are privately owned, and 14 units were designed for assisted living.

Residents are highly discouraged from owning an automobile. They can purchase one if they please, but a lack of parking and ample access to alternative transportation renders doing so impractical. If they wish to use a car, it is expected that they participate in a citywide car share. There is no on-site parking, except for a few spaces for residents and visitors. It is important to note that vehicle access to the community is granted for emergency and service vehicles. As well, a number of

Fig. 19.5 In the Slateford Green project in Edinburgh, Scotland, the developer created a place that illustrates the benefits of car-free living

parking spaces are located outside of the community but are exclusively for health and maintenance workers.

Many goods and services are available within proximity to Slateford Green. A childcare center is within the community and a primary school is 500 m (1640 ft) away. Recreation facilities are located nearby; these include a public park, football field and bowling yard—all within 1 km (0.62 miles). Shops, grocery stores, and a shopping center are in one of the surrounding neighborhoods. Several bus lines run along the roads bordering the community. Using the bus service, the city center can be reached within 10 min. The main transit station is located less than 3 km (1.86 miles) away.

References

Ben-Joseph E (1995) Changing the residential street scene: adapting the shared street (Woonerf) concept to the suburban environment. J Am Plann Assoc 61(4):504–515

Collarte N (2012) The Woonerf concept: rethinking a residential street in Sommerville. http://www.solaripedia.com/files/1325.pdf. Accessed 8 Mar 2020

Jacobsen PL, Racioppi F, Rutter H (2009) Who owns the roads? How motorised traffic discourages walking and bicycling. Inj Prev 15(6):369–373

Pucher J, Buehler R (2008) Making cycling irresistible: lessons from the Netherlands, Denmark and Germany. Transplant Rev 28(4):495–528

Steinberg L (2015) Woonerf: inclusive and livable Dutch street. https://www.humankind.city/post/woonerf-inclusive-and-livable-dutch-street. Accessed 08 Mar 2020

Chapter 20
Public Transit and Urban Design

Abstract Public transit is an essential feature of sustainable mobility networks. Such systems contribute to the reduction of greenhouse gas emissions and may prevent economically vulnerable people from having to own and maintain a vehicle, which is an unavoidable and draining investment for those in communities where there is limited, or no public transit. All modes of public transit must be an integral part of the planning process. To do so, the existing regional network must be considered, and public transit must be designed in conjunction with streets and paths in mind, while paying attention to local residents' needs. This chapter outlines urban design strategies for well-integrated public transit networks.

Keywords Multimodal Transit Networks · Transit networks · Transit stops · Walking · Cycling · Public transit · Bike sharing · Active transportation · Congestion · Car park controls · Street design · Light rail transit · Safe spaces · Transit stops

20.1 Designing Multimodal Transit Networks

Public transit may better serve communities by alleviating both negative environmental impacts through the reduction of greenhouse gas emissions, and economic stress placed on households having to purchase and maintain a private vehicle. For public transit to be successful, it must be well integrated in the existing urban fabric at a larger regional scale, as well as consider features, such as street design and road networks, that will maximize the volume of residents that will use public transit.

All modes of transportation which exist alongside public transit must be considered when planning the network. A multimodal approach to network planning is a way to ensure that the system is well integrated with other modes of mobility. Public transit should also be designed in harmony with existing roads and paths to synchronize the activities on them. It should be made accessible and safe for residents; design strategies that can be implemented to make this achievable will be outlined later in this chapter. Finally, the design of the transit stops is also an important feature to consider, as they directly influence the efficiency or convenience of public

A. Friedman, *Fundamentals of Sustainable Urban Design*,
https://doi.org/10.1007/978-3-030-60865-1_20

transit. Planning for them is therefore another element of public transit that may enable or disrupt the proper functioning of public transit networks.

Considering different modes of transportation, including active and public transit and how they are organized, is important so that some convenience and versatility may be offered to commuters. Multimodal planning, which refers to "planning that considers various modes (i.e., walking, cycling, public transit) and connections among modes" (Litman 2017), should therefore be implemented. Buses, subways, or trains are all different modes of public transit which must complement, rather than compete with each other; biking and walking are modes of active transportation which can be more accessible if they are incorporated in plans for public transportation. Furthermore, public transit is too often thought of in isolation from other modes of transportation, although these can be understood as being complementary to it.

Planning for multimodal public transportation networks does present certain challenges. One of the main challenges is that the demand for each mode of transportation is often difficult to predict. The capacities are not easy to estimate, as one-way travel may occur (King 2014). For instance, a rider could travel to a destination using the subway but return using bike sharing. Furthermore, a study by Stephen Krygsman et al. (2004) found access and egress to be the most challenging aspects of multimodal public transport, as waiting and transferring times make it especially discouraging for commuters. It is therefore recommended to coordinate between different modes of public transportation and to make them appealing to commuters.

Recent research has hierarchically arranged modes of transportation, to determine which mode should be given priority. By doing so, more efficient modes could be emphasized when allocating road space, as well as the funding and pricing of different modes (Litman 2017). Private motorized vehicles are on the bottom of the hierarchy, followed by carpools, taxis, service and freight vehicles, then public transportation, and finally cycling and walking at the very top (Litman 2017).

Research has shown that public transit is much more efficient when it is planned close to walking and cycling infrastructure (Fig. 20.1) (Gatien and Mas Baghaie 2019), which is why including modes of active transportation when planning public transit could add more value to the public transit. A study by Ugo Lachapelle and Robert Noland (2012) demonstrated that walking and public transit use may be supported by transit services and neighborhood destinations as complementary aspects to such modes of active transportation. Transit services must therefore be relevant to the needs of locals residing in neighboring and regional locations. Neighborhood destinations should also be in relation to transit services so that commuters are encouraged to walk or use public transit to reach key areas. What is important to consider, however, is that private motorized vehicles should also be considered when planning for an integrated multimodal public transit network, as these will be a part of systems. The idea is that people will drive and park near a regional public transit stop, take transit to their destination, and use their car upon return. Similarly, some may ride their bike and store it near a transit stop to use them at the end of the day (Fig. 20.2).

Fig. 20.1 Public transit is much more efficient when it is planned close to walking and cycling paths as is the case in Malmö, Sweden

Fig. 20.2 Storage depot for cyclists who may ride their bike and store it near a transit stop and reuse them at the end of the day as shown in Copenhagen, Denmark (left) and Perouges, France (right)

Hong Kong and Singapore are known to be exemplary cities when it comes to integrated multimodal transport planning (Litman 2017). The public transit share mode in Hong Kong is up to 90%, and available modes of transportation include railways, trams, buses, minibuses, ferries, and taxis. However, in addition to providing public transport, Hong Kong also relies on congestion and car park controls to diminish car usage (Luk and Olszewski 2003). Similarly, in Singapore, although the government is known to invest in public transport infrastructure in a consistent manner (Litman 2017), it also relies on high car ownership and usage costs as a means to discourage the use of private motorized vehicles (Luk and Olszewski 2003). Cars are therefore also a part of the multimodal public transit network, as they play their part in promoting the use of public transit.

20.2 Public Transit at the Street-Scale

Street design plays an important role in making all public transportation accessible, convenient and attractive. Streets occupy more than 40% of the city's public space and should be designed for individuals to easily and safely walk, cycle, or use public transit.

There are certain design strategies which urban areas can follow in order to prioritize and improve the service and delivery of active and public transit through the design of streets (NACTO 2016). As discussed in Chap. 18, path width can either promote or discourage the use of certain modes of transportation. As explored in Chap. 19, narrower streets have shown to slow down driving speeds, which makes it safer to walk and more time consuming to drive. In this setting, walking would be the preferred mode of transportation, which would naturally lead to more public transit use (Gatien and Mas Baghaie 2019). Lanes may be allocated to private motorized vehicles, buses, bikes, or parked cars which is why lane width should be assessed with existing space and should respond to all needs (NACTO 2016).

Lanes dedicated to buses, streetcars or on-street light rail transit should be built using the existing urban fabric (NACTO 2016). Dedicated transit lanes ensure that commuters save times, regardless of traffic congestion (Fig. 20.3). These lanes could be curbside or offset, meaning that the rightmost travel lane of streets where parking is usually permitted is instead dedicated to buses, streetcars, or on-street light rail transit. They could also be median lanes, meaning that lanes are placed at the centerline of multilane roadways, and these should be planned with the accessibility of transit stops in mind (NACTO 2016).

Attractive designs of public transit also make it more inviting and therefore more accessible for residents who use it. Public transit could be made more attractive by ensuring infrastructure is properly maintained and buses, metros, and trains remain clean and safe spaces, such as is the case in Singapore which is known for its successful public transit network (Fig. 20.4) (Litman 2017).

Fig. 20.3 Dedicated transit lanes ensure that commuters save times regardless of traffic congestion as this one in Melbourne, Australia

Fig. 20.4 Public transit could be made attractive by ensuring that the stops are properly maintained and safe such as is the case in Singapore

20.3 Designing Public Transit Stops

Transit stops should be well located and convenient to reach for commuters. The spacing, distance from homes and design of stops are critical to consider when planning public transit (Fig. 20.5). It is important to consider both the larger regional existing environment and local level features of the area surrounding the stops (NACTO 2016).

There are three common designs for bus stops based on where they are located: far side bus stops, nearside bus stops and midblock bus stops (NACTO 2016). Far side bus stops are the type preferred by designers, and are located after an intersection, driveway, or crosswalk (Fig. 20.6). They not only allow buses to move through

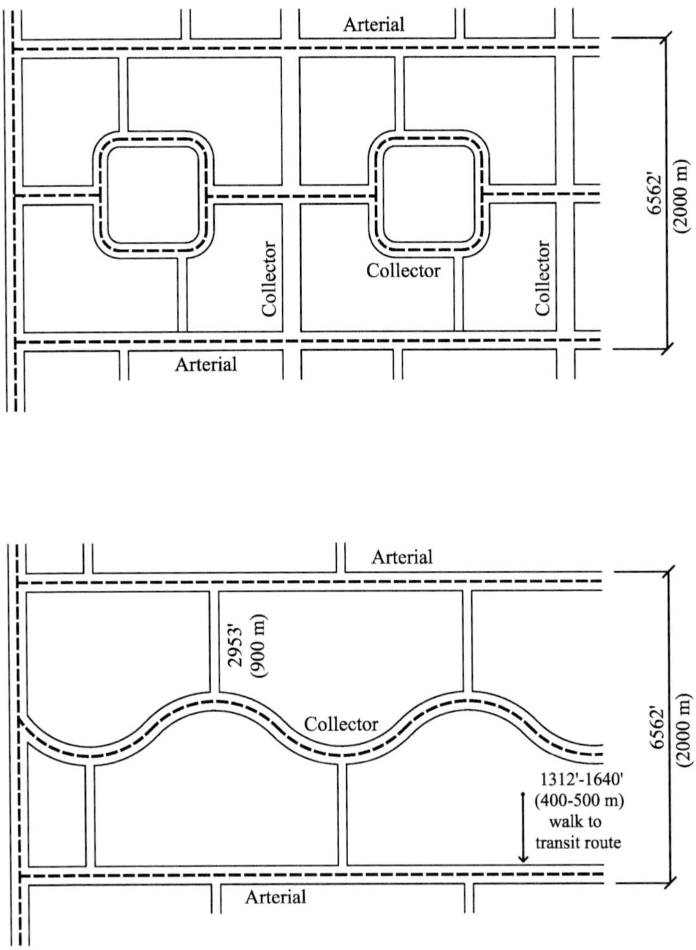

Fig. 20.5 The spacing and short distance of bus stops from homes are critical to consider when planning public transit

Fig. 20.6 A midblock streetcar stop in Melbourne, Australia

intersections in a more efficient way, but pedestrians can also safely pass a stationed bus (BC Transit 2018). Nearside bus stops are located before intersections and are generally only implemented to make specific transfers more accessible or when far side stops are not possible to design (BC Transit 2018). Midblock bus stops are only necessary when planning a bus stop nearside or far side of an intersection is not possible due to physical or environmental obstacles, or when the block is very long. In this case, a midblock crosswalk should be placed behind the bus stop, so that pedestrians are prohibited from walking in front of buses (BC Transit 2018).

Public transit stop should be placed near main activity points such as schools or shopping centers. Placing a convenience store near a bus shelter for instance would be a good strategy to make sure commuters are well served when they are waiting for the bus. Furthermore, locating transit stops near key amenities would also ensure that commuters are able to reach important destinations via public transportation.

Public transit stops should be designed to make waiting time as pleasant and convenient as possible (Fig. 20.7). Bus shelters should ensure that commuters are protected from the elements and serve as a community meeting point. For instance, transit shelters may be designed to be heated in colder climates or to provide adequate shade in hot climates. Benches, bus schedules, street maps of the

Fig. 20.7 Public transit stops should be designed to make waiting time as pleasant and convenient as this one in Berlin, Germany

Fig. 20.8 Timetable and maps of a regional transit system should all be included in a transit stops or shelters as the one shown in Copenhagen, Denmark (left) and Lyon, France (right)

surrounding neighborhood, and maps of the regional bus transit system should all be included near transit stops or shelters (Fig. 20.8).

Designing successful public transit networks requires several considerations. All modes of transportation must be considered when planning for a multimodal public transit network. The local existing urban fabric also has specific features that may influence whether public transit is accessible, efficient or convenient to residents. Finally, transit stops also play their part in making public transit successful in the local context and are important elements of urban design.

References

BC Transit (2018) BC Transit infrastructure design summary. https://www.bctransit.com/documents/1507213895398. Accessed 23 Mar 2020

Gatien A, Mas Baghaie A (2019) Improving active transportation and public transit integration: a guidebook for policy and planning. https://www.tcat.ca/wp- content/uploads/2019/06/Active-Transportation-and-Public-Transit-Integration-web-3.pdf. Accessed 23 Mar 2020

King D (2014) Three big challenges for planning multi-modal cities. https://www.citylab.com/design/2014/10/3-big-challenges-for-planning-multi-modal-cities/381254/. Accessed 23 Mar 2020

Krygsman S, Dijst M, Arentze T (2004) Multimodal public transport: an analysis of travel time elements and the interconnectivity ratio. Transp Policy 11(3):265–275

Lachapelle U, Noland RB (2012) Does the commute mode affect the frequency of walking behavior? The public transit link. Transp Policy 21:26–36

Litman T (2017) Introduction to multi-modal transportation planning: principles and practices. Victoria Transport Policy Institute, Victoria, BC

Luk N, Olszewski P (2003) Integrated public transport in Singapore and Hong Kong. Road Transp Res 12(4):41–51

National Association of City Transportation Officials (NACTO) (2016) Transit Street Design Guide. https://nacto.org/publication/transit-street-design-guide/. Accessed 23 Mar 2020

Chapter 21
Urban Design and Shared Transport

Abstract The need for a local response to climate change and easing traffic congestion is among the challenges that occupy the minds of many city planners and administrators. A solution embraced by cities worldwide has been the introduction of shared transport. This type of transport includes more conventional modes such as carpool networks or bikes and car sharing systems, as well as newer and more technological ones such as electric bikes or scooters. Shared transport stands out for its worldwide success, and it is often described as a product of the digital age, as most shared transport modes rely quite heavily on technology. This chapter investigates the urban integration of shared transport in cities while paying attention to the preferred location of stations and their design.

Keywords Shared transport · Electric bikes · Digital age · Public transportation · Affordability · Accessibility · Carpooling · Vanpooling · Multimodal transportation · Bike sharing · Geo-fencing technologies · Dockless system

21.1 The Growth of Shared Transport

Shared transport can be defined as a "demand-driven vehicle-sharing process where travelers share a mode of transportation, either at the same time (e.g., ride-sharing) or over time (e.g., bike sharing or car sharing)" (HERE Mobility 2020). This can be understood as a hybrid of public and private transportation, allowing users to share travel costs (HERE Mobility 2020). Most shared transport modes could not have been made possible without technological advances; technology is a precondition to using many shared mobility services by allowing individuals to request, track and purchase trips through their phones (APTA 2016).

Research has found that the presence of shared transport tends to decrease the use of private motorized vehicles, while encouraging the use of public transportation (Katzev 2003). This type of transport therefore not only provides environmental benefits by reducing pollution and congestion, but also social and economic benefits by increasing the affordability and accessibility of transportation. Cities have however shown concern with regards to the management of certain shared

A. Friedman, *Fundamentals of Sustainable Urban Design*,
https://doi.org/10.1007/978-3-030-60865-1_21

Fig. 21.1 Shared transport can include carpooling vehicles as is the case in the Grow Community in Bainbridge, Washington, USA

transport modes, particularly with electric bikes and scooters. Parking has been the central issue, as bikes and scooters tend to be randomly placed near sidewalks or other public spaces, acting as obstacles for pedestrians and making streets appear cluttered and disorderly.

Common shared transport modes include bikes, electric bikes (E-Bikes), electric scooters, and cars used for carpooling or vanpooling as is the case in the Grow Community in Bainbridge, Washington, USA (Fig. 21.1). Such shared transport modes have become quite popular in many urban areas; shared transport provides users with more flexibility for the route, time, choice of vehicle, and operator.

Research has also shown that shared transport may contribute to increased public transit use, promoting well-integrated multimodal transportation networks as discussed in Chap. 20. The American Public Transportation Association (APTA) conducted a study in order to explore challenges and opportunities related to "technology-enabled mobility services," referring to shared transport, and how public transit and these modes may be better integrated between each other.

Shared transport users were found to use public transit more, and shared modes were found to complement public transit. Furthermore, shared transport modes are growing in popularity which is why public transit agencies are collaborating with private companies to improve urban mobility for individuals (Fig. 21.2) (APTA 2016). The key findings support and encourage the need for more cooperation between public mobility providers who are responsible for public transit and private mobility providers who are often in charge of shared transport modes (APTA 2016).

A study by Christine Cheyne and Muhammad Imran (2015) undertaken in New Zealand showed how shared transport may offer great social and economic benefits

Fig. 21.2 In some cities like Rome, Italy, the local transit agency also manages the shared car system

to residents of small towns and larger urban areas by increasing transportation options for everyone including more vulnerable groups such as low-income households, the elderly, and people with disabilities. Shared transport allows users to travel to main centers for employment, education, health, or other services without necessarily having to own a vehicle. Furthermore, new digital platforms are heavily used by shared transport modes, which allows these modes of transport to be more flexible than most public transportation. This provides an incentive for transport-policymakers and planners to incorporate shared transport modes within transportation systems, as flexible transport is key to environmental sustainability and inclusivity (Cheyne and Imran 2015).

A shared transport mode which has become quite common in most cities is bike sharing programs. Bikes can be rented by customers at a variety of locations throughout a city. The bikes are rented using a self-service kiosk or phone-based application from one station and can be returned at any other station in the network. These programs have been successfully implemented in many cities around the world, improving access to active transport (see Chap. 19). Some bike sharing programs include electric bikes, providing a pleasant and efficient riding experience for users, as pedaling can be assisted by the motor, limiting the effort required from riders even on the steepest hills (Fig. 21.3) (Thompson and Kennan 2017). This improves the accessibility of shared transport to include seniors. Electric bikes offer a cheap and convenient transit option for individuals in dense urban areas through the use of technology. Urban areas have shown to greatly benefit from electric bikes

Fig. 21.3 Electric bikes or E-bikes, like this one in Copenhagen, Denmark offer a pleasant and efficient riding experience as pedaling does not require too much effort even on hills

by reducing pollution, congestion, and noise as the use of private vehicles decreases. However, a challenge which many cities have faced has been establishing regulations which fit within the urban context, and this has led to some cities having to ban electric bikes. Providing proper infrastructure such as designated bike lanes and secure bike parking has also been discussed as being key to ensure that electric bikes become more prevalent in cities in a controlled and safe way.

21.2 Strategies for Placing and Designing Stations

Planners should locate stations for shared transport modes in a way that ensures both order in the urban landscape and convenience for the users. Furthermore, because most shared transport modes are new, it has been difficult for cities to manage them and strategically place them within the urban landscape and the existing public transportation and road networks.

Local governments will necessarily have to provide designated parking spaces for small vehicles such as bikes or scooters within public space. While some cities allow individuals to leave their rented bike and scooters anywhere and have unrestricted or “free floating” parking, other cities require parked vehicles to be in the curb strip or furniture zone (NACTO 2018). Unrestricted parking allows users to leave their vehicles anywhere that is not within ADA-required sidewalk space, making point to point trips easier for users. However, this can quickly result in streets appearing cluttered and impede on other spaces such as sidewalks or driveways, as well as make access difficult for pedestrians (NACTO 2018).

Fig. 21.4 A car2go parking station in Vancouver, Canada

Enforcement is limited, as it can be difficult to inform users where the vehicles can be left or not (NACTO 2018). Furthermore, GPS and geo-fencing technologies prove to be limited and therefore make enforcing small vehicle parking through the use of data difficult to monitor. Most cities rely on spot-checks and reports of problems to monitor whether regulations are being respected. However, geo-fencing technologies may potentially ensure parking locations once they are further developed (Fig. 21.4) (NACTO 2018).

Parking stations for scooters and bikes have been requested by cities to promote order and decrease clutter in streets (Bliss 2019). Although most shared transport modes are "dockless" in nature (NACTO 2018), they can nevertheless be designed to be dock-based, and this is often the fact with bike share systems where bikes can be returned to a specific location and locked there (Thompson 2018). A solution has been offered by start-up company Swiftmile where solar powered docks charge scooters at the parking stations. These docks are also able to collect data about vehicle use and condition and are suitable for all scooter models (Bliss 2019). However, the aim would not necessarily be to make parking stations for all scooters, but rather to provide enough docking space for about 25% of all scooters, especially in areas with high pedestrian activity and limited sidewalk space, so that companies can still reap the economic benefits associated to a dockless system, as it allows them to invest into providing more scooters rather than incurring the costs of docking infrastructure (Fig. 21.5) (Bliss 2019).

Fig. 21.5 Several dockless scooters are parked in Lyon, France

Fig. 21.6 A bike rental station near a subway stop in Vienna, Austria

Some studies have looked at rental bikes specifically and found that locating parking stations near other public transit stations and in areas with lower traffic levels to encourage the use of shared transport mode as well as public transport was the best approach (Fig. 21.6). One of the main weaknesses of public transport is the access to public transport stops; locating rental bike stations near them would make public transport stops easier to reach, and therefore make individuals more inclined to use public transportation. For instance, locating flexible rental bikes at train stations was found to reduce car use, increase the number of train trips, and increase the use of bicycles for nonrecurrent trips in the Netherlands (Martens 2006).

Ensuring parking facilities are located somewhere that is safe for bicycle users is also something important to consider encouraging better access to public transportation stops. There exists a wide range of parking spaces for bikes in many Dutch, Danish, and German cities, and these parking spaces are located often near train stations (Pucher and Buehler 2008). Certain bike parking facilities are well designed. For instance, the main train station and bus terminal in Muenster, Germany, provides deluxe parking for up to 3300 bikes. The parking includes a ramp which connects the bike parking level to the street level, and which offers cyclists access to all train platforms (Pucher and Buehler 2008).

Many benefits are tied to shared transport modes, whether environmental, social, or economic, and it is therefore worthwhile to invest time and money into making them as successful as possible. In order to do so, such modes of transport must be integrated in existing public transit and road networks and should also be planned and designed in a way which considers the needs of both the users and surrounding environment.

References

American Public Transportation Association (APTA) (2016) Shared mobility and the transformation of public transit. https://www.apta.com/wp-content/uploads/Resources/resources/reportsandpublications/Documents/APTA-Shared-Mobility.pdf. Accessed 08 Apr 2020

Bliss L (2019) The hot new things in dockless electric scooters: docks. https://www.citylab.com/transportation/2019/03/electric-scooters-parking-charging-docks-lime-bird-lyft-spin/584332/. Accessed 08 Apr 2020

Cheyne C, Imran M (2015) Shared transport: reducing energy demand and enhancing transport options for residents of small towns. Energy Res Soc Sci 18:139–150

HERE Mobility (2020) Shared transport: moving toward a common goal. https://mobility.here.com/shared-transport-moving-towards-common-goal. Accessed 08 Apr 2020

Katzev R (2003) Car sharing: a new approach to urban transportation problems. Anal Soc Issues Public Policy 3(1):65–86

Martens K (2006) Promoting bike-and-ride: the Dutch experience. Transp Res A 41(4):326–338

National Association of City Transportation Officials (NACTO) (2018) Guidelines for the regulation and management of shared active transportation. https://nacto.org/wp-content/uploads/2018/07/NACTO-Shared-Active-Transportation-Guidelines.pdf. Accessed 08 Apr 2020

Pucher J, Buehler R (2008) Making cycling irresistible: lessons from the Netherlands, Denmark and Germany. Transplant Rev 28(4):495–528

Thompson H (2018) Combining E-bikes and bikeshares is urban alchemy: a winning transit solution for cities. https://www.forbes.com/sites/energyinnovation/2018/01/08/combining-e-bikes-and-bikeshares-is-urban-alchemy-a-winning-transit-solution-for-cities/#1ca9dbed3e09. Accessed 08 Apr 2020

Thompson H, Kennan H (2017) As transportation costs, emissions grow, electric bikes offer an efficient alternative. https://www.forbes.com/sites/energyinnovation/2017/07/06/as-transportation-costs-emissions-grow-electric-bikes-offer-an-efficient-lternative/#5a1a0e2a305a. Accessed 08 Apr 2020

Chapter 22
Accommodating Seniors and People with Reduced Mobility

Abstract In the past 50 years, the world has witnessed the proportion of senior citizens increase more rapidly than any other age group. In 1950, 205 million of the world population (approximately 2.5 billion) was aged 60 years or above. Half a century later, this figure had increased several-fold and is expected to reach two billion by 2050. It is therefore necessary to adapt urban environments to the needs of the aging population, and those with reduced mobility. As motor and mental abilities decline further among seniors, the need to pay attention to the design of urban features becomes paramount. This chapter discusses the design of urban environments for the aged and people with reduced mobility.

Keywords Seniors · Mental abilities · Reduced mobility · Inclusive · Independence and integration · Safe streets · Improved health · Wealth · Education · Motor output · Social isolation · Economic sustainability · Age-friendly communities · Wayfinding

22.1 The Scope and Challenges of Designing for Reduced Mobility

The United Nations predicted that a third of the population will be over 60 by 2050 (2006). The World Health Organization suggests that, up to one million people reach retirement age every month (2015). This change will pose an urban design challenge, since some older people face physical restrictions that make using everyday spaces more difficult. Designing cities that consider an aging population's needs requires a shift in design, so that urban areas are inclusive of older citizens.

As the population ages, urban environments will have to undergo changes to accommodate seniors. Aging populations tend to have certain common characteristics that dictate a specific set of mobility needs that must be met by planners. However, these characteristics may change depending on age, since aging is a process and there are a range of problems seniors are likely to face depending on their age category. It is nonetheless important for all needs to be addressed, and a shift in

A. Friedman, *Fundamentals of Sustainable Urban Design*,
https://doi.org/10.1007/978-3-030-60865-1_22

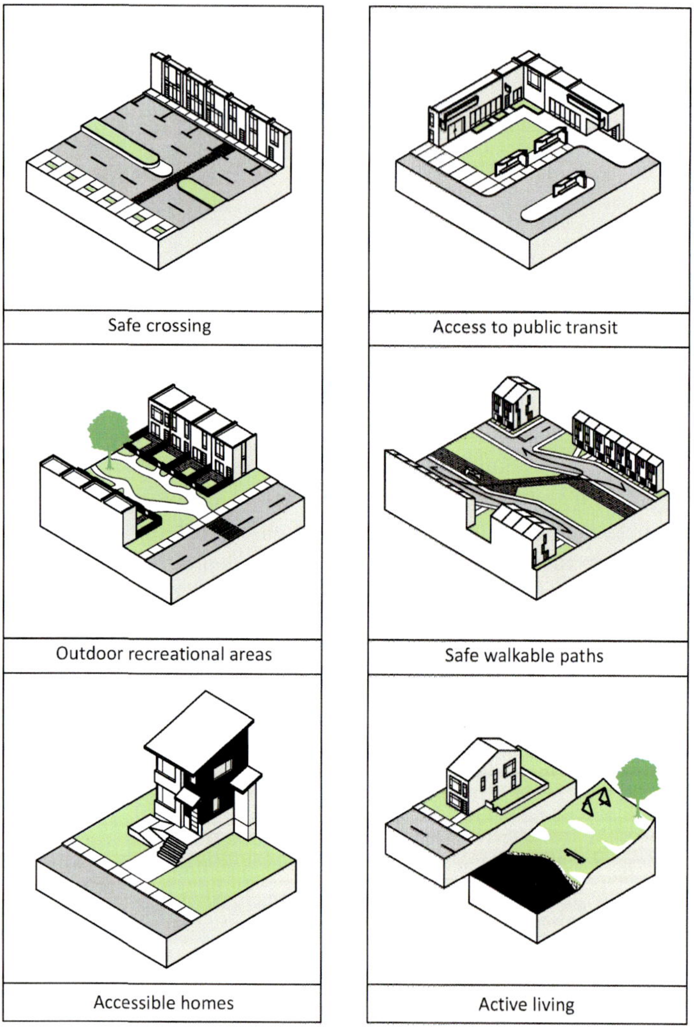

Fig. 22.1 Urban design aspects that accommodate seniors and people with reduced mobility

mindset is the first step toward doing so. Economic sustainability, independence and integration, forward thinking, and universal design are all concepts which should be considered when it comes to designing inclusive urban environments. Safety should be at the forefront of design initiatives, as it is key to making urban areas conducive to seniors' health and well-being. Planning while considering safe streets and sidewalks, adapted public transit and open spaces is important to ensure successful urban design that meets the needs of the aging population and people with reduced mobility (Fig. 22.1).

It is important to determine the profile of the aging population to find ways to accommodate their needs. Contemporary characteristics which define the aging population and distinguish it from the previous generations include computer literacy, improved health, wealth, education, extensive traveling, and familiarity with social media. Life expectancy has also significantly increased since the twentieth century, as data has shown that seniors who reach their 80s are very likely to live into their 90s.

Aging is a progress phenomenon with different phases in which seniors will not necessarily require the same needs. For instance, someone approaching 70 may experience the death of their peers or the loss of a spouse and may also experience the loss of sensory acuity (Howell 1972). People in their 80s will most likely experience different changes mostly related to their deteriorating health such as diminished motor output, age-related health problems or overall reduced physical mobility.

The aging population will therefore have a range of needs, not necessarily tied to their age (Howell 1972). Generally, seniors are likely to experience social isolation as their networks of friends and the number of social interactions diminish (Fig. 22.2). This may lead to decreased mental health or mood disorders such as depression, which is often untreated. Seniors may also feel financial stress, as statistics show that seniors spend most of their life savings in the last 2 years of their lives. Not only does the aging population require care for several years, as seniors are expected to live longer, but the different phases under which the aged will go through must also be considered. Long-term or reoccurring problems that seniors face should be avoided when possible and this can be done by setting up the proper structures and planning accordingly.

There are current challenges concerning the aging population that need to be considered and addressed. First, the way cities, communities, and the services found

Fig. 22.2 A public glass structure in Malmö, Sweden designed to accommodate seniors' gathering in all seasons

Fig. 22.3 In some suburban communities, streets are not designed to accommodate people with reduced mobility

Fig. 22.4 Poorly maintained sidewalks can put at risk the walking of seniors and people with reduces mobility and poor eyesight

within them are currently resourced, organized, and delivered often disadvantages older adults with chronic health problems (Sinha 2012). Another concern are suburban settings whose designs are not ideal for seniors. In some suburban communities, the roads are not designed to accommodate people with reduced mobility and in other places sidewalks do not exist or are poorly maintained (see Fig. 22.3 and 22.4). Many seniors also have no other choice but to live in nursing homes or require home care, which often results in financial strain. Adults have been providing an increasing percentage of financial assistance to their aging parents in the past years (MetLife 2011).

A shift of mindset is needed in order to design mobility in cities and communities in a way that better serves the aging population. Four main concepts should be kept in mind: economic sustainability, interdependence and integration, forward thinking, and universal design. When it comes to economic sustainability, it is important for investment into senior's independence and well-being to occur, as this will save government funds in the short and long run. Successful interdependence and integration are other factors to consider, as urban design must support seniors in their everyday lives for as long as possible in communities.

Community planning for seniors should be done correctly from the start, as retrofitting is more expensive; forward thinking is a crucial element for planning and designing cities to cater to the needs of seniors. Finally, the fourth concept is universal design. Universal design means that a space may be accessed, understood and used to the greatest possible extent, by persons of any age, size or ability, in the most independent and natural manner possible, in the widest possible range of situations, and without the need for adaptation or modification. It is by considering these four concepts that successful planning and designing of cities for the aging population will be achieved.

22.2 Urban Design Strategies

The World Health Organization defines age-friendly cities as "inclusive and accessible urban environments that promote active ageing" (WHO 2009). These communities necessarily include age-friendly homes designed in a way which makes living easier for older people. Safety is a key objective which should be incorporated in the design of streets, sidewalks, public transit, and open spaces in urban environments (Fig. 22.5) (see Chap. 18).

Safe streets are a main feature of age-friendly communities. Such streets should be car-free, including green buffers and one-way streets, have expanded medians,

Fig. 22.5 Well marked sidewalks and accessways for the visually impaired in Singapore

Fig. 22.6 Narrow, slow vehicular speed streets and well build sidewalks for wheelchair users create safe environments for seniors and people with reduced mobility as shown in Leiden, the Netherlands (left) and Langford, British Columbia, Canada (right)

surface changes at intersections, and slow traffic. Safe streets also include signaling such as traffic lights, prolonged street crossing times and both visual and audible alerts, so that seniors may comfortably and safely cross streets (Fig. 22.6).

Additionally, wayfinding solutions need to be provided along streets and public resting places. Wayfinding include unique landscaping, distinctive public art and bold colors that are visually embedded into the built environment to function as distinctive landmarks. Highly visible and recognizable features are important to include to help seniors orient themselves. Repetitive streetscapes and housing can make it difficult for seniors suffering from memory loss to get to shops and return home. Chapter 31 will outline artistic ways to implement wayfinding solutions in cities.

Sidewalks are also a key element of pedestrian landscapes in age-friendly urban areas. These should include seating of various sorts and have efficient snow and ice removal systems put in place, so that seniors may safely walk on sidewalks without the fear of getting injured throughout the year. Renewable resources have been used to heat sidewalks to melt the snow and ice; Aomori, Japan uses geothermal and solar systems while Reykjavik, Iceland uses heat collected from hot springs and recovered waste heat to keep sidewalks clear throughout the winter (Orkustofnun n.d.). Sidewalk surfaces should be easy to walk on so that all seniors, including those supported by walking aids, experience convenient and safe paths, free from obstacles.

As the number of fatal car crashes by drivers are highest among the older population, efforts should be made to encourage them to use public transportation. Features such as sheltered bus stops, automatic door openers, and adapted transportation (e.g., ramps at bus entrances for wheelchairs) are simple yet effective ways to make public transit more senior friendly and attractive to the aging population (Fig. 22.7).

Two other features which should be considered when it comes to safety are sightlines and entrapments in open spaces. Concerning sightlines, sudden changes of grade and blind corners should be minimized, while the visibility in high risk areas should be maximized. Entrapment spots should be eliminated or closed after operating areas, and opportunities should be provided to escape from entrapment spots or to find help. Entrapment spots should also generally be made more visible for the safety of all community members.

Fig. 22.7 Features such as automatic door openers (left), and ramps at bus stops for wheelchairs user (right) are effective ways to make public transit more senior use-friendly

Fig. 22.8 Having small parks with comfortable sitting (top left), community gardens (top right), fitness equipment in public places (bottom left) and organized activities, are important for seniors' health and social engagement

There are also opportunities to make open spaces inclusive and relevant to seniors. For instance, planning small parks, community gardens, and exercise places, may help do so, as these are places seniors are likely to be in (see Chap. 28). Furthermore, open spaces can also become important meeting places where the aged may socially interact by engaging in different activities such as dancing or group exercise (Fig. 22.8).

Designing urban areas in ways which are inclusive to the aging population by catering to their needs is necessary, as the world population is getting older. There are specific challenges which must be addressed, but these may be overcome if urban design is done successfully. Not only is urban design able to make urban landscapes relevant to the aging population, but it also has the potential to deal with the challenges facing it.

References

Howell SC (1972) Privacy as an expression of human territoriality. In: Palastan L, Carson D (eds) Spatial behavior of older people. University of Michigan Press, Ann Arbor

MetLife (2011) Market survey of long-term care costs. https://www.aarp.org/content/dam/aarp/livable-communities/old-learn/health/the-metlife-market-survey-of-nursing-home-assisted-living-adult-day-services-and-home-care-costs-2011-aarp.pdf. Accessed 11 May 2020

Orkustofnun (n.d.) Snow melting. https://nea.is/geothermal/direct-utilization/snow-melting/. Accessed 17 May 2020

Sinha S (2012) Living longer, living well: recommendations to inform a seniors strategy in Ontario. http://www.health.gov.on.ca/en/common/ministry/publications/reports/seniors_strategy/docs/seniors_strategy_report.pdf. Accessed 14 May 2020

United Nations (2006) Population aging 2006. https://www.un.org/en/development/desa/population/publications/pdf/ageing/wallchart-ageing2006.pdf. Accessed 11 May 2020

World eHealth Organization (WHO) (2009) WHO global network of age-friendly cities. https://www.who.int/ageing/Brochure-EnglishAFC9.pdf. Accessed 13 May 2020

World Health Organization (WHO) (2015) World report on aging and health. WHO Library Cataloguing. https://apps.who.int/iris/bitstream/handle/10665/186463/9789240694811_eng.pdf;jsessionid=D2CD5F07B0FA2DE0861F9681C920B12D?sequence=1. Accessed 13 May 2020

Chapter 23
Accessibility and Livability in Winter Cities

Abstract Winter conditions are challenging, as they make accessibility and livability harder in many areas in the northern hemisphere. The need to consider these conditions in urban design is important, as healthy lifestyles which include both physical activity and social interaction should carry on in urban areas challenged by harsh winter conditions. This chapter examines aspects and strategies that need to be considered in the design of such cities.

Keywords Physical activity · Social interaction · Outdoor spaces · Winter cities · Physical barriers · Urban design strategies · Microclimates · Covered sidewalks · Climate-responsive approaches · Exclusionary regulations · Climate sensitive design

23.1 Scope and Challenges

Cities should be designed to support accessibility and livability during harsh winter conditions. Snow, ice, wind, and cold temperatures make simple activities such as walking and cycling more challenging, reducing social interaction during the winter months. Finding solutions through urban design is therefore very important and may be done either by creating more indoor community spaces, or by rethinking outdoor spaces, making them accessible throughout the year. The idea of embracing winter, rather than escaping it, has become quite popular when designing winter cities in a way that addresses some of the challenges they face. The term "winter cities" has been given to cities found in the northern hemisphere which experience harsh winter conditions. Although there is no specific definition of winter cities, they can be described as having all or one of the following five conditions: they experience subzero temperatures, precipitation is mainly in the form of snow, sunshine and daylight are limited to a few hours per day, the climates show considerable seasonal variation, and these conditions occur every year for several months in succession (Chapman 2018; Pressman 2004). People are naturally less likely to engage in outdoor physical activity in winter cities; it is for this reason that a key

A. Friedman, *Fundamentals of Sustainable Urban Design*,
https://doi.org/10.1007/978-3-030-60865-1_23

Fig. 23.1 A key challenge in the urban design of winter cities is to find solutions to promote outdoor activity year-round like children play as this one in Westlock, Alberta, Canada

challenge in the urban design of winter cities is to find solutions to promote outdoor activity year-round such as children play (Fig. 23.1) (Chapman 2018).

Winter conditions generally include darkness, coldness, wetness and snow accumulation, challenging urban life (Stout et al. 2018). Such problems tend to decrease accessibility and livability in winter cities, which in turn can decrease the health and wellbeing of residents (Davies 2015). There is often an overall decline in mental health because people stay indoors, limiting social interactions, exposure to sunlight, and physical activity. Seniors or citizens with reduced mobility especially suffer from social isolation, as they are likely to withdraw from public spaces in the winter due to dangerous physical barriers such as ice and snowbanks (Stout et al. 2018). Physical health is also compromised in the winter, as individuals tend to stay more indoors and are less inclined to engage in active forms of mobility such as walking or cycling. However, urban design has proven useful when it comes to dealing with some of these challenges in winter cities.

The "winter cities movement" refers to places that actively seek to become more "appealing and functional in winter," first and foremost through physical interventions (Stout et al. 2018). The winter city movement is a "design and behavioural approach to improve the quality of life in northern settlements, to ensure that winter is better coped with and celebrated, not just endured" (Davies 2015). There are certain urban design strategies which reflect the winter city movement's purpose by turning winter conditions into attractive features.

23.2 Urban Design in Nordic Cities

Several planning and design interventions for winter cities can be organized as either pertaining to the visual environment, to human comfort, to protective urban devices and strategies, to recreation and leisure, or to transportation (Pressman 1996). When it comes to the visual environment, the idea of using ice as art, bold colors on buildings' facades, illuminating dark spaces, incorporating urban furniture and civic embellishment are all ways to make the urban landscape more attractive and welcoming during the winter months (Fig. 23.2) (Pressman 1996). Promoting comfort involves having a better understanding of how microclimates function and using landscaping to reduce discomfort. Included in protective urban devices and strategies are above-grade protection such as skywalk systems and pedestrian bridges, below-grade protection such as underground pedestrian networks, and art-grade protection such as colonnade, covered sidewalks and canopies which may be linked to above-grade and below-grade protective systems (Fig. 23.3). Other protec-

Fig. 23.2 When it comes to the visual environment, the idea of using ice as art (right) and bold colors on buildings' facades (left) are understood as ways to make the urban landscape more attractive, engaging and welcoming during the winter months

Fig. 23.3 Included in protective urban devices of winter cities are above-grade protection such as covered sidewalks like the one shown on the left in Jackson Hall, Wyoming, US and pedestrian bridges over streets like the one shown on the right in Calgary, Alberta, Canada

tive measures include pavement heating, multiuse buildings, retractable roofs, and pedestrian or car-free zones. Interventions concerning recreation and leisure include implementing ski-trail networks, winter-oriented outdoor amenities, carnivals and festivals, and indoor gardens. Improving transit by shifting schedules in response to seasonal demands, or including heated shelters for instance, are key elements toward making transportation more convenient and accessible during the winter months (Pressman 1996).

Policies that address the challenges winter cities face must be translated into physical form if they are to have any effect, which is why urban design is such an important part of dealing with accessibility and livability issues in winter cities (Pressman 1996). There needs to be climate-responsive approaches and energy-efficient policies to be implemented both at the micro- and mesoscales for them to effectively improve livability in winter cities (Pressman 1996). The microlevel involves physical alterations and focuses on building design, while the meso level includes larger areas such as specific neighborhoods or commercial streets within cities, which is where opportunities for the most improvement of public outdoor spaces may be done (Stout et al. 2018). Although physical interventions are likely to vary in scale, these should be split into two categories: to provide residents with warm, safe and enjoyable indoor spaces, or comfortable outdoor spaces designed in accordance with the climate (Fig. 23.4) (Stout et al. 2018).

Fig. 23.4 Physical intervention in a winter city should provide residents with enjoyable spaces and activities like this indoor beach in the West Edmonton Mall in Edmonton, Alberta, Canada

Fig. 23.5 A common example where people can walk, socialize, and shop during winter in an enclosed space are shopping centers

Designing enclosed and heated pedestrian spaces so that people may walk, socialize, and shop is one of the ways to respond to winter conditions. The most common examples of such enclosed spaces are grade-separated pedestrian systems and shopping centers that are attached to them (Fig. 23.5) (Stout et al. 2018). This interconnection of buildings is achieved through planning regulations which fit new and existing buildings together in a way that connects them to the above or below ground systems (Davies 2015). Examples of grade-separated pedestrian systems include underground connections and plazas, such as Toronto's PATH network and Montreal's La Ville Souterraine (Davies 2015). However, research shows that these enclosed spaces are not the best urban design solution when it comes to dealing with the challenges facing winter cities. Such spaces tend to be exclusive, as they are mainly intended for commercial purposes, rendering them into semipublic spaces subject to private control and exclusionary regulations (Stout et al. 2018). It is therefore important to focus on existing public spaces and to design them in ways that make them comfortable, attractive, and inclusive for all individuals during the winter month.

23.3 Designing Outdoor Spaces and Promoting Active Transportation

Another more inclusive approach to dealing with winter conditions is to value outdoor public space and reenvision how this space may be designed and used to make the best of the winter months. Climate sensitive design, paired with more attempts to activate public spaces during the winter months, can result in winter cities being able to successfully cope with some of the challenges they face due to winter conditions by embracing winter rather than avoiding it.

Climate sensitive design is a very important concept when it comes to the urban design of winter cities and involves the combination of climatology and urban design (Stout et al. 2018). Protecting people from the elements can be done effectively by creating microclimates. Problems associated with the cold, snow, and darkness can be addressed by creating attractive microclimates and protecting pedestrian from winds through landscaping and the design of buildings, by orienting public spaces so that they feature sun, and by sheltering pedestrians from snow and rain through the use of overhead structures (Stout et al. 2018). Decreasing the amount of shade from buildings and avoiding the creation of wind tunnels along streets by orienting prevailing winds across the major wind directions is one way to decrease snow accumulation (Fig. 23.6) (Davies 2015). Furthermore, public spaces should be redesigned so that they include solar traps where low angle winter sun is found, sheltering people from the wind and shade, making the spaces more attractive year-round (Davies 2015). Including street furniture and vegetation to protect from prevailing winds is also a way to make public spaces more attractive and comfortable during the winter months. However, it is also important to consider the aesthetic appeal of physical interventions, to make certain areas or amenities appear more attractive and welcoming. For instance, using bright, bold, and colorful lighting may encourage people to gather in key outdoor spaces (Fig. 23.7) (Davies 2015).

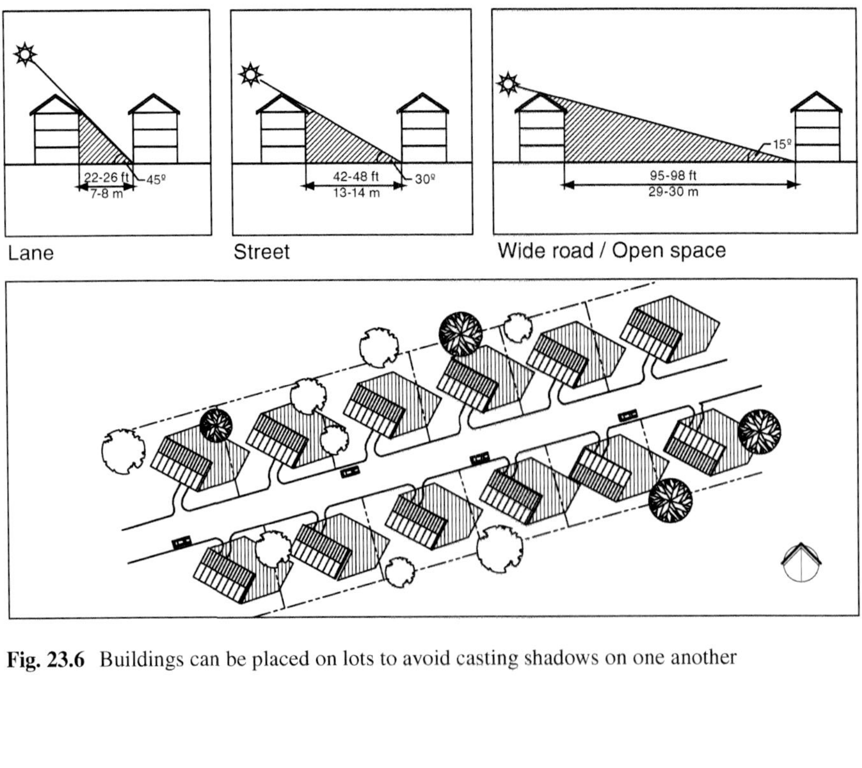

Fig. 23.6 Buildings can be placed on lots to avoid casting shadows on one another

Fig. 23.7 Using bright, bold colors on buildings like the ones shown in Copenhagen, Denmark (left) and lights like the one shown in Weslock, Alberta, Canada can encourage people to gather in key outdoor spaces during winter

Fig. 23.8 Accommodating users of public transit by installing enclosed bus shelters will encourage them to use the system

It is also important to accommodate public transit users and encourage soft mobility and active modes of transportation (Fig. 23.8). How existing infrastructure is managed is key to dealing with some of the problems winter cities face, as these should be made accessible for use during the winter months as well. For instance, one of the main factors affecting winter cycling rates is how well the city can maintain its bike network throughout the winter months (Jaffe 2016). Copenhagen is known to continue maintenance on its biking infrastructure even during the winter by prioritizing frequent snow removal on bike lanes (Jaffe 2016). Adapting pavements and roads to winter conditions is also important, in order to improve and maintain circulation patterns during the winter months. Frequent snow removal and placing fences alongside circulation paths to reduce snow accumulation on these surfaces are some examples of coping strategies.

Promoting the use of public transit during winter months may be done by making transit systems work more effectively and by providing an overall more comfortable experience for commuters. For instance, bus or rail shelters should be heated, wait times should be minimized by more frequent services, and public transit should be well integrated so that key transit destinations are connected to avoid outdoor waiting (Davies 2015).

Creating winter festivals, seasonal outdoor markets, and opportunities for recreational activities in the winter has also proven successful in terms of making public spaces accessible and livable during the winter months. For instance, Christmas markets in Germany are a good example of how outdoor spaces can be made vibrant and attractive, contributing to both socially and economically dynamic urban areas during the winter. These markets are events that are highly anticipated by people and that successfully make use of public space (Stout et al. 2018). Another example is reflected in Edmonton's WinterCity Strategy which emphasizes the importance of creating opportunities for people to engage in a variety of outdoor activities during the winter months, designed to celebrate winter and turn snow and ice into positive features rather than negative ones (Davies 2015). Edmonton, Canada, is known for its winter festivals, such as the Silver Skate Festival which attracted more than 50,000 people in 2019 despite temperatures averaging around −24 ° C (−11 ° F) during the 10 days of the event (Reith 2019). Edmonton's winter festivals and events include a variety of activities such as skating, skiing, or toboggan hills which are offered in welcoming and cheerful atmospheres, encouraging people to get outside, to stay active, and to socially interact despite the low temperatures (WinterCity Strategy 2018).

Accessibility and livability do not need to be compromised in winter cities during the winter months, as there exist many urban design strategies to cope with some of the challenges linked to difficult winter conditions. The main focus should not be to avoid winter by creating indoor enclosed spaces, but to take advantage of the already existing public places as well as pavements, roads and public networks, and adapting them to winter conditions. It is only a matter of re-envisioning spaces in order to make them accessible for use and attractive even during the winter months, so that people may stay active, healthy, and socially connected.

References

Chapman D (2018) Urban design of winter cities: winter season connectivity for soft mobility. Dissertation, Lulea University of Technology

Davies WKD (2015) Winter cities. In: Davies WKD (ed) Theme cities: solutions for urban roblems. Springer, Dordrecht, pp 277–310

Jaffe E (2016) How to keep cyclists riding even in the frigid snowy winter. https://www.citylab.com/transportation/2016/01/winter-bike-riding-seasonal-cycling/426960/. Accessed 05 May 2020

Pressman NEP (1996) Sustainable winter cities: future directions for planning, policy and design. Atmos Env 30(3):521–529

Pressman, N (2004) Shaping cities for winter: climatic comfort and sustainable design. *Prince George, Winter Cities Association*

Keith T (2019) Embracing and Edmonton winter takes layers of clothing—and a leap of faith. https://www.cbc.ca/news/canada/edmonton/edmonton-winter-embrace-1.5401015. Accessed 05 May 2020

Stout M, Collins D, Stadler SL, Soans R, Sanborn E, Summers RJ (2018) "Celebrated, not just endured": rethinking winter cities. Geogr Compass 12(8):1–12

WinterCity Strategy (2018) For the love of winter: strategy for transforming Edmonton into a world-leading winter city. https://www.edmonton.ca/city_government/documents/PDF/COE-WinterCity-Love-Winter-Summary-Report.pdf. Accessed 05 May 2020

Part IV
Public and Green Open Spaces

Chapter 24
Open Spaces as an Urban System

Abstract According to a 2018 United Nations report, by 2050, more than 7.7 billion people will be living in urban areas. One can predict that this rapid rate of urbanization, if uncontrolled, will reduce the quality of open spaces in cities. In response to the dilemmas posed by these forces, this chapter explores the urban design concepts which will comprise the foundation of a resilient open space system for urban landscapes. The chapter outlines four types of open spaces: the green belt arrangement, separate patches, and hierarchy configuration; the chapter illustrates how open spaces must be connected to form a coherent and resilient urban landscape.

Keywords Green belt · Separate patches · Hierarchy configuration · Nodes · Links · Corridors · Wellbeing · Urban density · Existing natural features · Connected green networks · Landscaping · Rainwater runoff · Playground

24.1 The Importance of Open Spaces

Though urban development tends to focus primarily on building areas zoned for construction, it is often the creation of open spaces which are as important to successful community planning. The inclusion and seamless integration of open spaces within urban areas is vital to the functioning of a city as a whole. Within this context, the term "open spaces" refers specifically to areas left unbuilt, while sectors adjacent to them are developed. They are either public or private, exist on the rural to urban spectrum, vary in scale, and preserve or create green infrastructure (Regional Open Space Strategy 2017). Furthermore, these spaces can be connected by patches, nodes, links, and corridors to form a system of such spaces (Kong et al. 2010). Therefore, urban open spaces are relied upon by urban dwellers for mental health, physical activity as well as leisure and are key to the functioning of a city. While their presence is fundamental to the resident's wellbeing, they are often compromised by the changing urban realities.

Rapid, uncoordinated urbanization creates a growing disparity between constructed and open spaces as cities focus on housing citizens without a concern for

A. Friedman, *Fundamentals of Sustainable Urban Design*,
https://doi.org/10.1007/978-3-030-60865-1_24

Fig. 24.1 In some nations, such as China, high-density living resulted in the diminishment of open spaces

their wellbeing once settled. New high-density urban spaces can negatively affect the lives of their residents as they confine them to dense areas with minimal access to open space (Fig. 24.1). Despite the historic tendency to create open spaces during the preliminary phase of city building, the modern desire for housing has diminished the preservation of open spaces. However, urban resilience should be prioritized above urban density; the concept of resilience in the urban environment denotes a place's capacity to absorb, adapt, and respond to changes that the urban system meets over time.

24.2 Typology of Open Spaces in Cities

To ensure the proper introduction and functioning of open spaces as a system, it is crucial to select the appropriate type to implement in each context. This section examines the four prevailing urban space types—greenbelt, separate patches, the hierarchy configuration, and composite—and discusses patterns of their usage (Fig. 24.2).

The green belt arrangement is a long and linear park space that is located around the perimeter of an urban area. It also functions as an urban growth boundary, as shown in Fig. 24.3, to restrict or direct the flow of metropolitan growth and encourage biodiversity by creating a place for wildlife mobility and prosperity (Bilgili and Gökyer 2012).

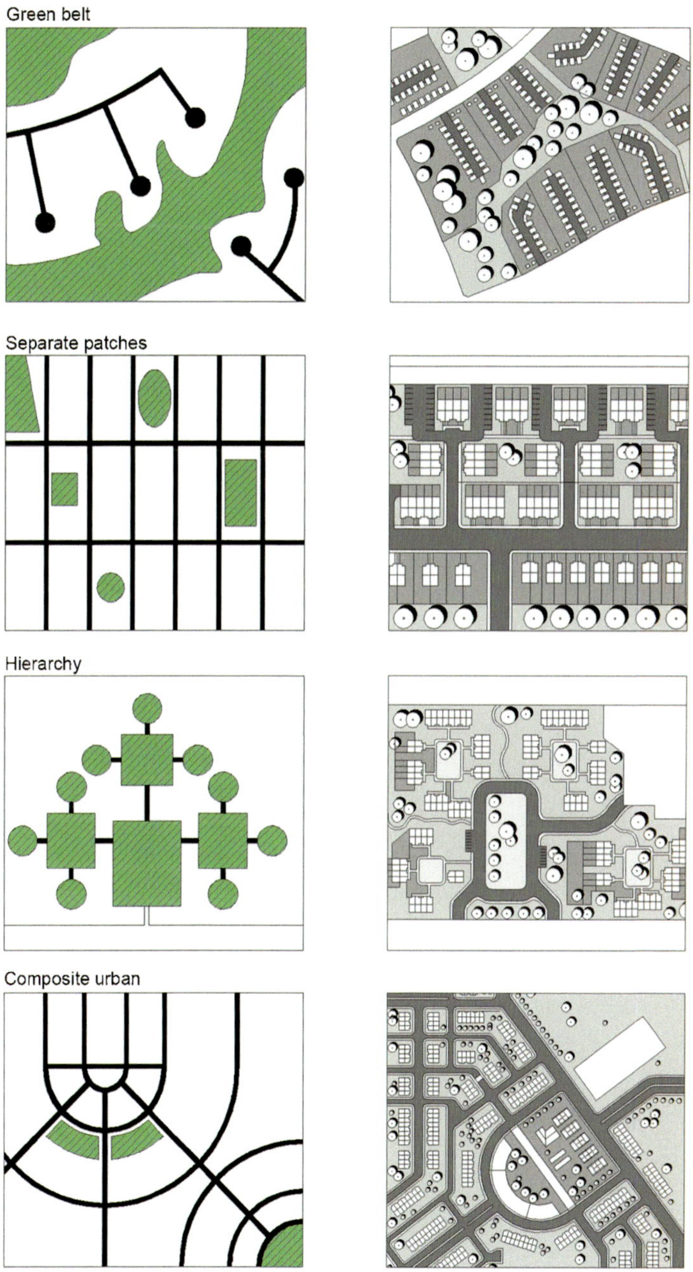

Fig. 24.2 Types of open space distribution can be categorized as green belt, aspartate patches, hierarchy and composite urban

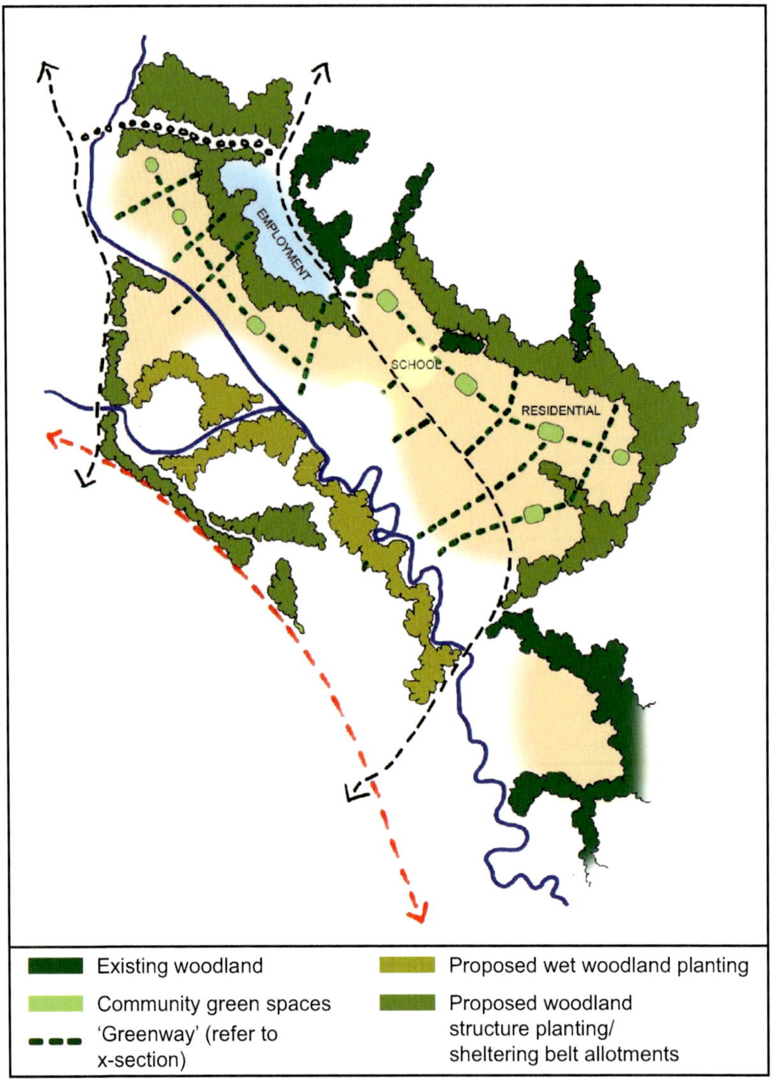

Fig. 24.3 The Scottish town of Inverurie integrated existing patches of green areas and new planting in its plan

Separate patches are fragmented open spaces that are dispersed without forming a cohesive pattern or establishing continuity. They have enclosed designs which make them appropriate to a gridiron-style urban area and function as an intimate and inviting gathering place in which to linger (Fig. 24.4).

The hierarchy configuration is flexible and is comprised of several connected open spaces of difference sizes. In practice, this manifests as a combination of large green spaces that link to medium-sized neighborhood parks and small-scale pocket parks by green corridors and pathways (Fig. 24.5). The diversity of scales inherent

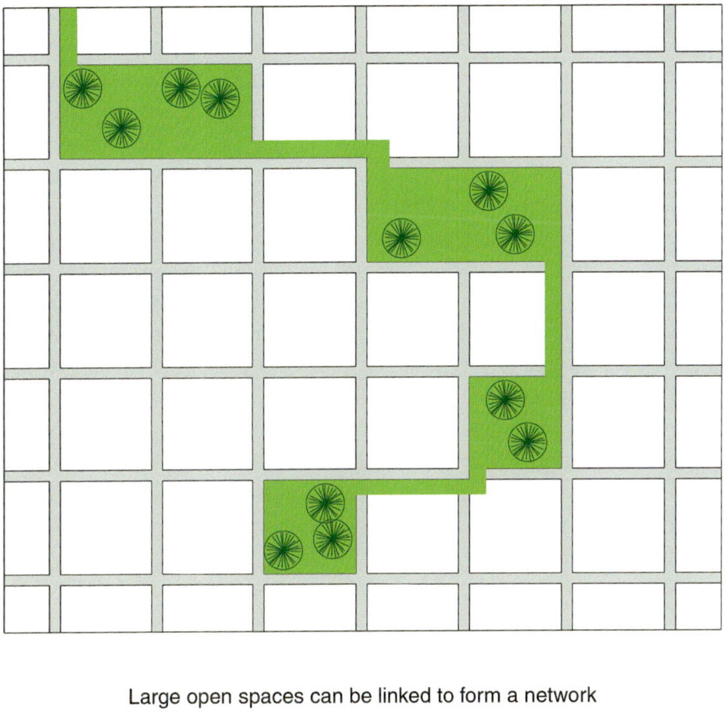

Fig. 24.4 Separate patches are fragmented open spaces that are dispersed without forming a cohesive pattern or establishing continuity. They have enclosed designs which make them appropriate to a gridiron-style urban area

Fig. 24.5 Small open spaces in residential areas known as Pocket Parks like this one in Montreal, Quebec, Canada belong to the Separate Patches category

to hierarchical open spaces allows for a variety of activities. The plethora of recreational opportunities that can take place within such a spatial arrangement enable it to cater to all age groups and make it a conducive type for dense urban areas.

Composite urban open spaces are enclosed and form geometric configurations. The enclosed and rigid nature of its design can hinder the site's incorporation of existing natural features such as slopes and vegetation. However, by placing a composite urban open space near large housing developments, it can function as a main artery that connects residential areas to an existing network of public, communal spaces.

Each of the above types of open spaces perform an array of important functions and have useful qualities depending on its urban context. A comprehensive study of the local environment and dynamic is suggested to determine which type of open space would be most appropriate for each urban environment. Moreover, open space systems work best when its selection and design is prioritized as a step in the preliminary stages of planning a new community. By selecting the type of open space before residential or commercial development commences, planners can ensure that these crucial areas are not neglected. Instead the spaces can be thoughtfully planned, executed, and incorporated into the city.

24.3 Strategies for Forming Open Space Systems

Research indicates that the ecological value of connected green networks is significantly greater than the sum of their individual green space parts (Mahmoud and El-Sayed 2011). Therefore, this section explores the urban design practices that enable various open spaces to cohesively form a resilient and interconnected system.

Firstly, the success of a public open space is often dependent on its proximity to residential areas (Fig. 24.6) (Friedman 2012). Moreover, an essential aspect of a functional system of open spaces is the pedestrian and cyclist-friendly paths that link them together into a coherent network. Walkways provide inhabitants with a web-like spatial configuration which they can navigate at their leisure. Furthermore, choosing to place pedestrian pathways on the predominantly sunny sides of open spaces is an effective way to encourage their usage during colder months of the year— an important technique for designing winter cities.

A key feature of resilient open spaces is their continuous utility throughout the year in the face of adverse weather conditions. Awnings and canopies over paths can reduce weather's harsh impacts by mitigating wind, snow, rain, and ice. Additionally, elevated fly shelters that fasten to trees or surrounding infrastructure diminish the severity of snow cover, allowing walkways and leisure areas to be user-friendly even for those with mobility restrictions. Moreover, the implementation of shelters throughout an open space system provides pedestrians with a shield from weather elements. Shelters that feature inviting furniture inside and a transparent material coverage above to allow sunlight in during the day are valuable for users. Alternatively, the installation of shelter roofs with passive solar design can effec-

Fig. 24.6 The success of a public open space is often dependent on its being part of residential areas

tively provide lighting in the evening. Furthermore, designing the infrastructure's exterior walls to use dark colors rather than light ones, promotes the absorption of heat over an extended period to keep the shelter's interior warm. In contrast, to accommodate hot weather, landscape design should prioritize vegetation cover to provide shaded areas to mitigate the urban heat island effect and improve air quality (Zhang et al. 2019).

In addition to specifically weather-conscious urban designs, landscaping plays an important role in the success of an open space network by mitigating weather effects and delineating land uses. Currently, insufficient sewer network capacities results in sewer overflows, which affect human health, and risk aquatic life (Rathnayake and Anwar 2019). In response, the use of rain gardens is an effective way to manage storm water (Fig. 24.7).

Vegetation in rain gardens filter the rainwater runoff, recharging groundwater aquifers. The adaptability of size, location, and cost-effectiveness of rain gardens allow them to be implemented in various locations throughout an open space network. Moreover, they effectively collect rainfall all year, provide aesthetic value in the summer, and function as snow storage during the winter.

Moreover, landscaping efforts can utilize vegetation as a gentle buffer to differentiate between various land-use areas. Strategic plant placement can block prevailing winds to create sun traps which make open spaces warmer throughout the year.

Fig. 24.7 Rain gardens are an effective way to manage storm water like the ones shown here in Singapore

In both cases, vegetation can include hedges, trees, hills, and more. Spaces can be separated by function using landscaping, ensuring the provision of a variety of activities that maximizes the amount of available space for users of all ages.

Open spaces are most successful when they prioritize and maximize the amount of people who use them without barriers to age, physical ability, or social status. For example, protective rubber floors are conducive to active and passive recreation, as they allow a variety of activities to take place such as ball games, farmers markets, and skating rinks in the winter. While it may seem contradictory, designing open space networks that attract people of all ages and interests requires constructing spaces that are activity-specific within a larger open space system. In doing so, designers may attract citizens to participate in large open space hubs by drawing them to a single open space within a network. For example, consider a children's playground that is adjacent to a community garden. The playground should be in an area without heavy vehicular traffic with connecting pedestrian linkages to make it accessible to those of all ages and physical abilities (Fig. 24.8). Additionally, the playground should be within seeing and hearing distance of an adult-friendly open space, in this case, an urban garden. These considerations ensure that both the playground and garden perform useful individual roles and combine to function productively as part of the open space network.

It is imperative that urban designers conduct a careful examination of the site's existing conditions prior to the implementation of appropriate strategies. Furthermore, it is important to utilize strategies that contribute to the interconnected nature of open space systems and whose adaptability is applicable to the weather, as well as variability in mobility and intergenerational usage. Ultimately, these diverse innovations each play an important individual role within a critically interconnected network of open spaces to promote social and ecological sustainability.

Fig. 24.8 Playgrounds need to be in areas without heavy vehicular traffic and well connected to make them accessible to people of all ages and physical abilities like the one shown here in Vienna, Austria (top left), Budapest, Hungary (top right) and the two bottom ones in Berlin, Germany

References

Bilgili BC, Gökyer E (2012) Urban green space system planning, landscape planning. Ecol Appl 22(1):349–360

Friedman A (2012) Open spaces. In: Town and terraced housing: for affordability and sustainability. Routledge, Abingdon, pp 156–179

Kong F, Yin H, Nakagoshi N, Zong Y (2010) Urban green space network development for biodiversity conservation: identification based on graph theory and gravity modeling. Landsc Urban Plan 95(1–2):16–27

Mahmoud AHA, El-Sayed MA (2011) Development of sustainable urban green areas in Egyptian new cities: the case of El-Sadat City. Landsc Urban Plan 101(2):157–170

Rathnayake U, Anwar AF (2019) Dynamic control of urban sewer systems to reduce combined sewer overflows and their adverse impacts. J Hydrol 579:124150

Regional Open Space Strategy (Ross) (2017) University of Washington UW Green Futures Research & Design Lab, Seattle, WA

United Nations (2019) World urbanization prospects: the 2018 revision. United Nations, New York

Zhang Z, Meerow S, Newell JP, Lindquist M (2019) Enhancing landscape connectivity through multifunctional green infrastructure corridor modeling and design. Urban Urban Green 38:305–317

Chapter 25
Integrating Existing Natural Features

Abstract By 2040, it is predicted that approximately $60 trillion USD will be spent financing new construction in cities globally (zu Ermgassen et al., One Earth 1:305–315, 2019). The future and present rates of hyper-urbanization can be attributed to rapid population growth. Consequently, the amount of built-infrastructure required to accommodate this growth puts at risk existing natural assets. In response to this unfortunate reality, this chapter explores urban design concepts that enable successful incorporation of existing natural features into development to conserve flora and fauna.

Keywords Existing natural features · Ecological footprint · Ecosystems · Psychological attributes · Soil and rock formation · Native wildlife · Seasonal variability · Native species · Green networks · Runoff drainage · Paved surfaces · Flora

25.1 The Importance of Integrating Existing Natural Features

In 2010, the United Nations outlined three urgent challenges: improve the quality of life in cities, reduce cities' ecological footprint, and ensure cities' resilience to climate change. One of the solutions to these three goals lies in conservation and integration of existing natural features into new developments (Pauleit et al. 2017).

As discussed in Chap. 24, excessive urbanization poses a great threat to the presence of natural open spaces within cities. Human activity leads to conflict between building and environmental conservation, as developers often see the integration of natural assets to be a barrier to development. Consequently, green open spaces are gradually disappearing from the day-to-day experience of urban dwellers (Fig. 25.1).

The prevailing car-centric urbanization is known to threaten natural areas which worsens air quality for example. Living in overcrowded cities which lack natural features is greatly associated with insufficient physical activity, exposure to anthropogenic environmental hazards, and chronic stress (Braubach et al. 2017). Therefore, hyper-urbanization is both harmful to the natural environment and to humans.

A. Friedman, *Fundamentals of Sustainable Urban Design*,
https://doi.org/10.1007/978-3-030-60865-1_25

Fig. 25.1 Clearing urban forest to make room for residential development in Langford, British Columbia, Canada

The integration of natural features is important because it improves the quality of life for urban dwellers and preserves nature. On the ecological level, a successful ecosystem is crucially dependent on the preservation of all-natural aspects. Traditionally the practice of conservation focused on protecting flora and fauna and begins by taking stock of existing assets (Fig. 25.2). However, new progressive approaches aim to ensure the entire ecosystems' ability to thrive and expand across spatial and temporal boundaries (Van der Biest et al. 2020). Nature conservation is thus crucial as it ensures the availability of basic human necessities including potable water, clean air, and uncontaminated food.

Socially, the integration of natural features can be understood as a component of public health as it can reduce stress (Barton and Rogerson 2017). Moreover, exposure to existing natural features benefits physical health by promoting physical activity, reducing risks of cardiovascular disease, obesity, and diabetes (Braubach et al. 2017). The following section explores effective strategies that integrate existing natural features into open space systems.

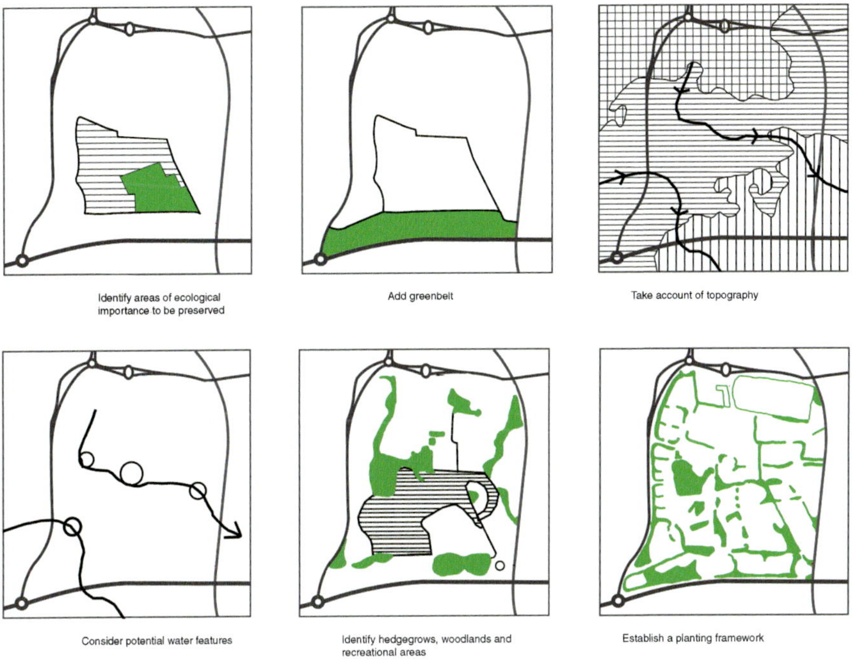

Fig. 25.2 Conservation that focuses on protecting flora and fauna begins by taking stock of existing natural assets

25.2 Typology of Natural Features

To ensure the proper integration of existing natural features in development, it is fundamental to understand the various types of natural features to develop the appropriate conservation practices. This section examines six prevailing types of natural features. Their relevant design integrations will be explored below.

Water provides immense value to open areas. Bodies of water enhance open spaces by providing visual, audial, tactual, and psychological attributes (Fig. 25.3). Moreover, the ecological benefits of a body of water to an open space includes replenishing groundwater, controlling floods, supporting wildlife, and helping spaces effectively adapt to climate change (Bindu and Mohamed 2016).

Topographic features are valuable to open areas; the consideration and inclusion of topographic features make open spaces visually unique and functionally interactive. Additionally, varying levels of terrain provide a natural drainage system, which minimizes the need for built-infrastructure.

A thorough understanding of the area's soil and rock formation is crucial to open space design. Designers can use this valuable information to create spaces that intelligently integrate the rock and soil's natural qualities. Therefore, planners ought to consider three main factors to soil conservation: the permeability of the soil, level of bedrock, and the water retention by the soil and vegetation.

Fig. 25.3 Bodies of water enhance open spaces by providing visual, audial, tactual, and psychological attributes

Vegetation is an extremely important consideration when integrating existing features into an open space. Designers need to plan with the intention of preserving as much existing vegetation as possible to mitigate the urban heat island effect, filter air, and reduce runoff (Fig. 25.4). Vegetation native to the area should be used because it supports native wildlife.

Exposure to sunlight encourages the use of an open space throughout the year. It improves mental health outcomes, increases alertness of users, and sustains spaces' vegetation. Careful analysis of sunlight patterns is recommended to design the space accordingly.

Wind in an open space is a valuable feature that has meaningful effects throughout the year. The strategic incorporation of breezes in relation to existing vegetation enables wind-control throughout the seasons and ensures usage during the warmer months of the year.

25.3 Urban Design Methods of Integration of Nature

The integration of existing natural features in a new open space is an effective way to enhance the area and minimize development costs. This section explores the urban design practices that enable successful integration of existing natural elements into an open space system.

Fig. 25.4 Preserving as much existing vegetation as possible mitigate Urban Heat Island, filter air, and reduce rainwater runoff

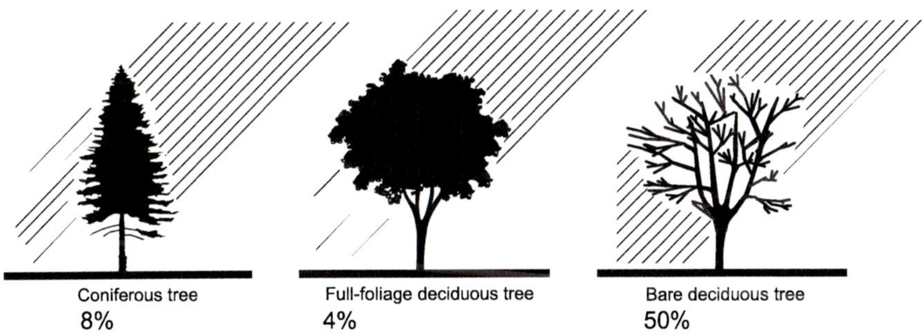

Fig. 25.5 Some tree species allow sunlight to penetrate through their sparse branches during winter and others not

Locating open spaces in strategic proximity to existing vegetation is a useful way to provide natural shade and ventilation while controlling sunlight. During the design process, it is essential to consider how shade, air flow, and sunlight can be incorporated into the open space and to consider their seasonal variability. Firstly, existing trees are useful to control wind flow, as they can function as shields for oncoming winds during the colder months of the year. In contrast, during the summer months, existing vegetation provides shade to minimize solar gain and reduce temperature. Moreover, orienting open spaces so that trees lie on its southern edge ensures optimal shaded area. Additionally, trees allow sunlight to penetrate through their branches during winter (Fig. 25.5). Beyond just locating open spaces in rela-

Fig. 25.6 Well-designed pathways have gentle curves which force users to stay alert and oblige cyclists to reduce their speed

tion to existing vegetation, it is also important to consider techniques to conserve a place's wildlife and native species.

Therefore, the design process must ensure the careful preservation of flora and fauna. If the ground must be excavated or paved over, designers need to protect the branches and intricate root systems that lie beneath the grounds surface. Creating new green spaces that connect to existing green networks is an effective way to create space while preserving existing natural features (Hagen and Stiles 2010). These interconnected systems allow fauna to move throughout the urban area and colonize surrounding natural habitats (Uslu and Shakouri 2013). In practice, pedestrian paths are an essential aspect of these web-like configurations.

As expressed in Chap. 24, pedestrian and bike paths are vital to the successful functioning of open spaces. Well-designed pathways have gentle curves which force users to stay alert and oblige cyclists to reduce their speed (Fig. 25.6). Their curved nature allows the passage to weave into local surroundings and existing flora without disruption. With that in mind, when mapping the placement of paths, designers need to ensure they do not cover trees' principal roots or locate paths too close to the trunks of mature trees. If paths are not carefully planned, their placement can reduce roots' access to water and make trees unstable. To accommodate rain and runoff drainage, paved surfaces should slope toward grass to funnel water into the soil or a rain garden where it can be productively absorbed. The construction of carefully planned paths which connect open spaces without harming existing flora is essential to open space system design.

Adapting to an open space's topography during the design process is also crucially important. If proper considerations are not made prior to construction, permanent damage to the local ecosystem can occur. Once terrain is fundamentally altered or removed it is challenging and costly to restore. Areas with varying levels of terrain are suitable for open spaces because they require minimal built-in. For example, natural inclines function as drainage systems. However, despite their advantages, designers must also consider that topographically complex spaces reduce mobility for users. The implementation of accessibility features like wheelchair- ramps and handrails can provide appropriate mobility solutions. Beyond simple external topographic features, designers must also evaluate complex local soil and rock formations.

Designing open spaces to preserve existing soil and rock formations is crucial (Fig. 25.7). There are four planning guidelines that should be considered to protect natural soil and rock quality. First, soil that is native to that area should be kept onsite and minimal amounts of offsite infill should be introduced. Second, cutting and filling the terrain should be avoided to minimize the disruption of the existing structural composition of the soil. Third, avoid the grading of properties to emphasize natural and existing drainage patterns. Finally, conserve existing vegetation

Fig. 25.7 Designing open spaces while preserving and integrating existing soil and rock formations is a paramount to the integrity of a landscape planning

including groundcover during the building process. Otherwise, the loss of ground cover can reduce the soil's ability to absorb water and prevent erosion.

Open spaces that are adjacent to waterways are advised to design the area with the water feature as the primary focus. Open spaces with bodies of water benefit from the tempering effects of evaporation. Lining the edge of bodies of water with rain gardens is an effective way to protect and enhance bodies of water. In addition to introducing more native plants to the area, the rain garden draws attention to the body of water and collects overflow caused by rainfall.

Boardwalks as a sustainable alternative to footpaths; are conducive to high visitor rates while simultaneously preserving their surrounding environment. Their benefits include their ability to: decrease erosion, enable the mobility of wildlife underneath, not disturb runoff or bodies of water, adapt to topography, protect rare flora or fauna, and restore vulnerable ecosystems. In designing boardwalks, important factors to consider the area's existing flora, fauna, and soil. It is essential to integrate existing vegetation between board walk planks (Fig. 25.8). To do so, designers must consider the needs of existing vegetation; the greater amount of space between planks, the more sunlight and water penetrate the vegetation underneath.

Boardwalk height is an essential consideration. Elevated boardwalks are optimal to provide wildlife and vegetation adequate space. Boardwalk height can be determined by assessing dynamic ecological processes such as wind transpiration of sand, determining the projected height of native flora and considering the extent of solar radiation that the area's soil requires. When implementing the urban design

Fig. 25.8 It is highly recommended to integrate existing vegetation between boardwalk planks rather than uproot it

strategies highlighted in this section, it is important to consider that the effectiveness of the solutions depends on the specific conditions of each site.

This chapter explored how urban design can address the reduction of biodiversity in cities caused by hyper-urbanization to establish resilience. It is demonstrated that the inclusion of existing natural features in urban open spaces provides sustainable vibrancy to spaces as it avoids significant alterations to the landscape. Moving forward, there are guidelines that should be considered in designing open spaces to incorporate existing natural features. First, sustainable materials must be sourced to build infrastructure. Second, the design process should encompass a bottom-up approach to integrating nature. This means designers should plan by prioritizing existing natural features above all else. Third, built-infrastructure should not only integrate but also highlight the existing natural features of the space. These key natural components can be incorporated and emphasized to encourage open-space-centered ecotourism.

References

Barton J, Rogerson M (2017) The importance of greenspace for mental health. Br J Psych Int 14(4):79–81

Bindu CA, Mohamed AR (2016) Water bodies as a catalyst to growth and development-the case of Kodungallur town, Kerala. Proc Technol 24:1790–1800

Braubach M, Egorov A, Mudu P, Wolf T, Thompson CW, Martuzzi M (2017) Effects of urban green space on environmental health, equity and resilience. In: Kabisch N, Korn H, Stadler J, Bonn A (eds) Nature-based solutions to climate change adaptation in urban areas. Springer, Cham, pp 187–205

Hagen K, Stiles R (2010) Contribution of landscape design to changing urban climate conditions. In: Müller N, Werner P, Kelcey JG (eds) Urban biodiversity and design. Blackwell, London, pp 572–592

Pauleit S, Zölch T, Hansen R, Randrup TB, van den Bosch CK (2017) Nature-based solutions and climate change–four shades of green. In: Kabisch N, Korn H, Stadler J, Bonn A (eds) Nature-based solutions to climate change adaptation in urban areas. Springer, Cham, pp 29–49

Uslu A, Shakouri N (2013) Urban landscape design and biodiversity. In: Özyavuz M (ed) Advances in landscape architecture. Namık Kemal University, Turkey

Van der Biest K, Meire P, Schellekens T, D'hondt B, Bonte D, Vanagt T, Ysebaert T (2020) Aligning biodiversity conservation and ecosystem services in spatial planning: focus on ecosystem processes. Sci Total Environ 712:136350

Chapter 26
Urban Design for Biodiversity

Abstract Biodiversity, the degree of variation of life forms within a given species, natural system, or a place, ensures the functioning of ecosystems and establishes multilayered benefits to human health and wellbeing (Collins et al., Land Use Policy 64:114–123, 2017). However, development that is prioritized over the environment threatens biodiversity. This chapter explores urban design concepts that enable successful conservation and creation of natural areas to enhance biodiversity. To do so, the chapter first explores the importance of diverse flora and fauna to an urban environment. It proceeds to examine the fundamental principles of planning for biodiversity and presents urban design solutions to retool cities.

Keywords Diverse flora and fauna · Urban environment · Biodiversity · Endangering species · Biotic homogenization · Range of species · Genetic variation · Ecosystem qualities · Biological interactions · Ecosystem cycles · Sensitive Urban Design (BSUD) · Urban heat island effect · Green roofs

26.1 The Importance of Biodiversity

Between 2000 and 2030 a projected 1.2 million sq. km (0.75 million sq. miles) of land will be urbanized globally, which will be the fastest expansion of the built environment in history (zu Ermgassen et al. 2019). Consequently, we are currently experiencing an array of negative impacts including habitat loss and fragmentation; resource unavailability; the introduction of invasive species; the modification of natural disturbance regimes; increased levels of chemical, light, and noise pollution; and alterations to local climates due to the urban heat island effect (Fig. 26.1). Broadly speaking, these consequences are responsible for endangering species, biotic homogenization, and lowering genetic diversity (Garrard et al. 2018). To counteract the negative effects of large-scale urbanization, we ought to prioritize conscious planning and implement urban designs that promote urban biodiversity.

A place's biodiversity is defined by the range of species, genetic variation, and ecosystem qualities associated with the abiotic components of an area including landscape features, drainage systems, and climate (Swingland 2013). Historically, it

A. Friedman, *Fundamentals of Sustainable Urban Design*,
https://doi.org/10.1007/978-3-030-60865-1_26

Fig. 26.1 Large expanse of roads and parking lots like the one shown in Melbourne, Australia contributes to Urban Heat Island

Fig. 26.2 Connected patches of green open spaces can contribute to a place's biodiversity like this one designed by the author for Langford, British Columbia, Canada

was believed that nature and development were two opposing forces that could not coexist. However, today this myth has been disproved, as we now understand that urban design must play a crucial role in the sustainable preservation and creation of biodiversity (Fig. 26.2). With every passing year, the preservation of biodiversity is more crucial as it becomes increasingly complex to recreate the ecosystems we destroy (Parris et al. 2018).

26.2 Fundamentals of Biodiversity

Biodiversity is commonly understood as the variety of life on Earth. According to urban ecologists, biodiversity is specifically conceived of as consisting of three intersecting forces: connectivity, cycles, and biological interactions (Parris et al. 2018).

Connectivity between biodiverse areas enables the mobility of animals and the propagules of fungi and plants across the urban landscape. A species' ability to relocate themselves is critical for the preservation of genetic diversity and distinct ecological communities. Additionally, the mobility of animals and propagules support the presence of metapopulations. Metapopulations are communities that are geographically separate yet connected with each other by dispersing themselves to rescue near-extinct populations and support the recolonization of vacant habitats. Connectivity conservation promotes protecting natural areas and connecting preserved areas to create a larger network from a series of isolated units (Perkl 2016).

Cycles are essential to maintaining biodiversity as water, nutrients, and energy cycles sustain ecosystems. Currently, prevailing urbanization practices inhibit the cyclical-nature of biodiversity due to the widespread use of impervious surfaces. Impervious surfaces are materials that are not porous and do not absorb water such as paved roads, rooftops, and concrete footpaths. These features generate great volumes of runoff during periods of rainfall which disrupt natural ecosystem cycles.

Biological interactions including competition for resources, symbiosis, and pollination are crucial to the identity of that location. For example, pollination is vital for the sustenance of native plant diversity. Pollinators include native bees, butterflies, wasps, and nectar-feeding birds. Currently, hyper-urbanization disrupts these pollinators in a variety of ways, subsequently reducing pollination due to urbanization and resulting in the extinction of plant species.

26.3 General Principles of Planning for Biodiversity

Despite the obvious advantages of designing biodiverse open spaces described above, it is still not common practice to implement these factors. Planning open spaces to promote biodiversity is difficult as planners must resist the practice of transforming biodiverse, rich, open spaces into built places. Based on contributions by ecologist Georgia Garrard, the following section explores six principles of planning for biodiversity (2018).

Planners can begin their approach to conservation by establishing the "scale of biodiversity" following a study of existing conditions as was done by the author in Stony Plain, Alberta, Canada (Fig. 26.3). Biodiversity should be planned for at the city level, rather than isolated neighborhood pockets. Planners can work in accordance with the Biodiversity-Sensitive Urban Design (BSUD) framework. The framework offers a guide for how to provide net benefit results for native species

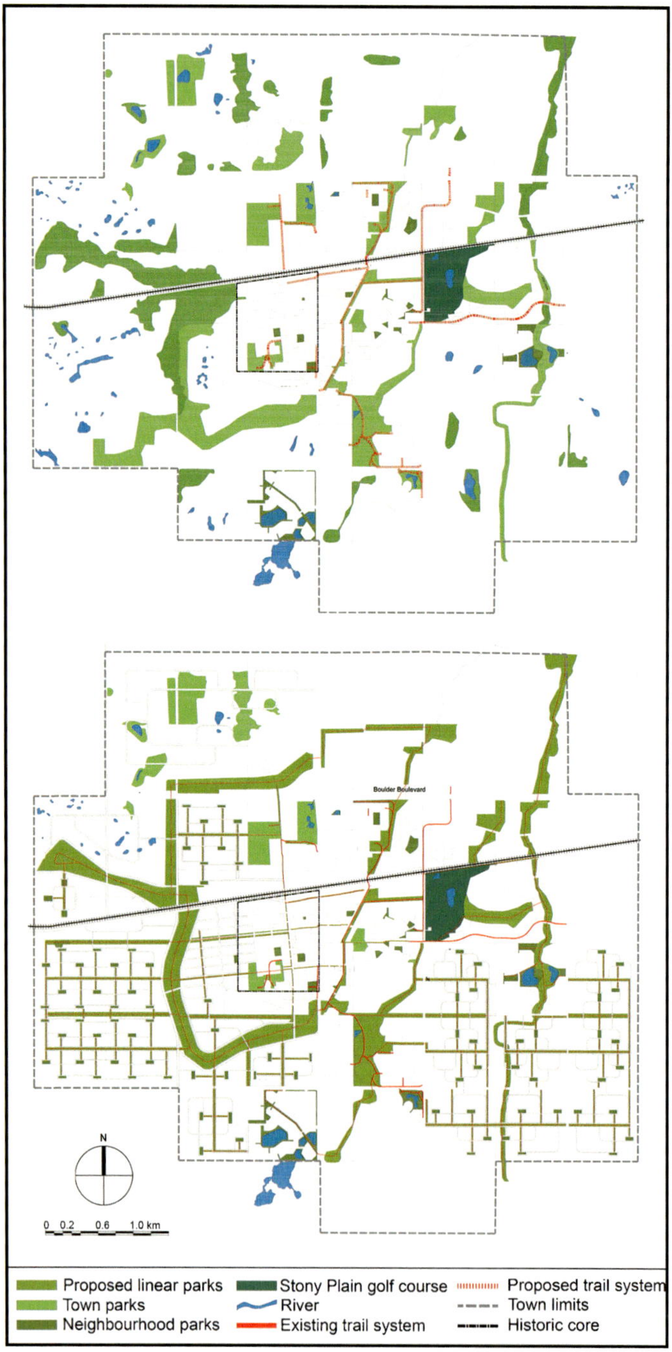

Fig. 26.3 Planners can begin their approach to conservation by establishing a scale of biodiversity following a study of existing conditions as was done by the author in Stony Plain, Alberta, Canada

and ecosystems by providing the necessary habitat and food resources (Parris et al. 2018).

Additionally, it is important to maintain existing habitats. New developments should be located in areas of low ecological value to avoid habitat loss. During the development process, planners must retain, protect, and incorporate existing vegetation. Planners should also work to create new habitats by introducing native plant species and adding green infrastructure.

Furthermore, it is crucial to facilitate dispersal during the planning process. In this vein, corridors can be constructed on both private and public land to promote mobility, implement infrastructure for animals and establish habitat connectivity.

Next, the planning process must minimize threats from anthropogenic disturbances. Anthropogenic disturbances are human caused impacts such as high volumes of runoff and nutrient loads as well as light pollution. To address issues such as stormwater runoff, plans should be designed in accordance with the principles of Low Impact Development (LID) (see Sect. 26.4.3). LID optimizes built infrastructure to sustainably coexist with the natural features of the surrounding area (Zimmer et al. 2012).

Lastly, it is important for planners to carefully consider the local context. Standardized urban design practices can sacrifice qualities that are unique to an area. Therefore, it is important to review landscape characteristics prior to making land-use decisions.

26.4 Retooling Cities for Biodiversity

There are a variety of urban design practices that enable the successful conservation and promotion of biodiversity into an open space system. This section draws upon the work of architect Manso and Castro-Gomes (2015) to examine the proper implementation of green roofs, green walls, storm water management, and the reduction of night light pollution.

The practice of covering buildings with vegetation greatly contributes to urban biodiversity as it supports storm water management, filters air, reduces hot temperatures, and mitigates the urban heat island effect. Green roofs and green walls are practical solutions to provide wildlife corridors in urban space and while creating a visually appealing space. When designing green roofs and green walls, there are four important aspects to consider: patch size, quality, abundance, and isolation. Moreover, planners must assess their implementations using a 3-dimensional green ecological network (Mayrand and Clergeau 2018).

Fig. 26.4 Green roofs actively promote biodiversity landscape connectivity in an effective way

26.4.1 Green Roofs

Green roofs actively promote biodiversity landscape connectivity in a meaningful way (Fig. 26.4). In their design, it is important to consider building height, native flora and fauna, and the construction process. Implementations that consider these factors influence the type and diversity of species able to use a green roof.

To ensure the inclusion of all wildlife, green roofs should be installed on buildings that are five stories or less (Mayrand and Clergeau 2018). Green roofs on buildings that are higher than five stories have greater exposure to wind and solar radiation, which reduces the quality of the habitat. Moreover, species with low mobility, such as carabids and spiders, are excluded from the habitat and taxa that are able to climb up may face difficulty returning to ground level safely (Lepczyk et al. 2017). Therefore, one should build green roofs at an appropriate height to ensure greater species access and to avoid becoming an ecological trap (MacIvor 2014).

26.4.2 Green Walls

Green walls play a similarly important role in urban biodiversity conservation. However, green walls are more effective because the surface area of wall facades are often double that of a buildings' roof area (Manso and Castro-Gomes 2015). Green walls refer to any form of vegetated vertical surface (Fig. 26.5). They are essentially vertical corridors that allow fewer mobile species to easily travel from ground level to green roofs. Additionally, green walls' assemblage of flora is beneficial to biodiversity. Their vegetation functions as a habitat for invertebrates and provides nutrients, nesting, and shelter for urban ornithology (Collins et al. 2017).

There are two types of green walls: green facades and living wall systems (LWS). Green facades have climbing or hanging plants directly attached to the wall and are either direct or indirect. Traditional direct green facades have plants directly grow-

Fig. 26.5 Green walls refer to any form of vegetated vertical surface. They can be on a home's exterior like the one on the left in Los Angeles, California, US or in a form of mobile unit like the one in Stuttgart, Germany on the right

ing onto the wall while indirect green facades install supporting structures to stabilize the vegetation. Indirect green facades can be further divided as either continuous guides or modular trellises. Continuous guides are green facades with one continuous support structure, whereas modular trellises consist of multiple support systems assembled to form a greater surface. Alternatively, LWS are characterized by their ability to provide rapid coverage of high buildings in a uniform fashion, adapt to various types of buildings, and integrate a wider variety of plant species compared to green facades. Similar to indirect green facades, LWS are either modular or continuous.

To ensure efficiency when installing, maintaining green walls, it is important to consider supporting elements, irrigation, and sensors. Supporting elements are frame-like vertical structures that climbing plants attach themselves to. They are either modular or continuous and anchor the vegetation's weight to stabilize the plant against wind, rain, and snow. To protect the wall from humidity, supporting frames have semipermeable, root proof screens stapled to their bases. Modular living wall systems are composed of several interlocking parts of lightweight materials such as plastic or metal sheets. The supporting structures of modular living walls take form as trays, vessels, planter tiles, or flexible bags. Trays and vessels are suspended from frames that are connected to the vertical surface.

Vegetation must have low irrigation needs and adapt to local weather conditions. For modular green facades and living wall systems, designers must implement an irrigation system to provide sufficient water to plants. To do so, tubes are installed at the top of the wall and connect the water flow to the central irrigation system. Irrigation tubes and connectors are made of rubber, plastic or silicone. Moreover, designers must install gutters at the base of the green wall systems to recover, store, and reuse water in the irrigation system. Plants that are native to the region are ideal as they require minimal irrigation.

Ongoing maintenance of living walls is best done by installing sensors. The sensors gather data about irrigation time and weather conditions such as quantity of rainfall, humidity, and atmospheric pressure. This informs landscapers about vegetation needs. Green walls are a sustainable solution to promote biodiversity. To maximize green patch size, designers should consider greening the roof and walls of the same building (Mayrand and Clergeau 2018).

26.4.3 Retention and Usage of Stormwater to Enhance Biodiversity

The retention and sustainable usage of stormwater is an effective way to enhance biodiversity. In practice, management of stormwater is achieved by designing in accordance with the principles of Low Impact Development (LID). Recall, the LID framework outlines a variety of methods to implement and promote biodiversity such as the effective management of stormwater (Fig. 26.6).

Fig. 26.6 Vegetation grow in canals fed by rainwater in Västra Hamman, near Malmö Sweden

Fig. 26.7 permeable interlocking pavement is a viable option for the ground cover which contribute indirectly to biodiversity like the one here in Los Angeles, California, US

As we explored in Sect. 17.3, permeable pavement is a viable option for ground cover which contributes indirectly to biodiversity. Permeable pavement is a porous surface composed of an open pore material, concrete and an underlying stone reservoir (Selbig and Buer 2018). This solution is adaptable to parking lots, low-traffic roads, sidewalks, and driveways. This design provides aesthetic benefits and supports stormwater management (Fig. 26.7).

26.4.4 Reduction of Night-Time Light Pollution

Lastly, planners must reduce light pollution where possible since artificial light interferes with animals' circadian rhythms, sleeping patterns, and navigation abilities (Parris et al. 2018; Macgregor et al. 2019).

Light regime, type, and distance are crucially tied to the seed count in the pollination process of pollinators. Light emitted at night is categorized into two types:

full-night lighting and partial-night lighting. Partial-night lighting regimes require lights to be switched off or dimmed at certain times in the night, whereas full-night lighting systems stipulate that lights are kept on and not dimmed throughout the night. Additionally, there are two types of lights to consider: high-pressure sodium (HPS) lights and light-emitting diodes (LEDs). Research suggests that pollination is maximized under partial-night lighting in tandem with high-pressure sodium (HPS) lights. While LED is less energy intensive than HPS, HPS is the better choice for natural environments, as it supports biodiversity. Therefore, planners should implement HPS lights in tandem with a partial-night lighting regime to manage the amount of energy consumed and preserve biodiversity. This solution saves energy, money, and supports nocturnal wildlife.

Evaluating the range of light that a fixture emits is an equally important consideration in reducing light pollution. The height and distance of a lamp from a biodiversity-rich area significantly determines the extent of light pollution. The further biodiversity-rich areas are located from lights, the better their pollination and thus the greater their yielded seed mass. Therefore, planners must determine the emitted light radius of lamps and ensure that it does not overlap with biodiversity-rich areas. With proper attention paid to light regime, type, and distance, humans can minimize the damaging effects of their urban lights on their natural neighbors.

This chapter explored how urban design can promote urban biodiversity to counteract the ecosystem loss caused by hyper-urbanization. Moving forward there are several guidelines that must be adhered to when designing open spaces to promote biodiversity. First, it is essential to remember that appropriate management of green infrastructure is equally as important as designing to promote biodiversity. Second, designers should prioritize greening existing built infrastructure to contribute to wildlife connectivity. An important first step in this process is the implementation of green roofs and green walls on buildings in urban areas with appropriate characteristics. Finally, when selecting vegetation types for these urban design implementations, designers must first inform themselves about native flora and fauna. A planners' proper consideration of these three factors is crucial to designing sustainable green infrastructure that promotes urban biodiversity by catering to the local wildlife context.

References

Collins R, Schaafsma M, Hudson MD (2017) The value of green walls to urban biodiversity. Land Use Policy 64:114–123

Garrard GE, Williams NS, Mata L, Thomas J, Bekessy SA (2018) Biodiversity sensitive urban design. Conserv Lett 11(2):e12411

Lepczyk CA, Aronson MFJ, Evans KL, Goddard MA, Lerman SB, MacIvor JS (2017) Biodiversity in the city: fundamental questions for understanding the ecology of urban green spaces for biodiversity conservation. Bioscience 67(9):799–807

Macgregor CJ, Pocock MJO, Fox R, Evans DM (2019) Effects of street lighting technologies on the success and quality of pollination in a nocturnally pollinated plant. Ecosphere 10(1):e02550

MacIvor JS (2014) Building height matters: nesting activity of bees and wasps on vegetated roofs. Israel Journal of Ecology & Evolution 62(1–2):88–96
Manso M, Castro-Gomes J (2015) Green Wall systems: a review of their characteristics. Renew Sust Energ Rev 41:863–871
Mayrand F, Clergeau P (2018) Green roofs and green walls for biodiversity conservation: a contribution to urban connectivity? Sustainability 10(4):985
Parris KM, Amati M, Bekessy SA, Dagenais D, Fryd O, Hahs AK, Hes D, Imberger SJ, Livesley SJ, Marshall AJ, Rhodes JR (2018) The seven lamps of planning for biodiversity in the city. Cities 83:44–53
Perkl RM (2016) Geodesigning landscape linkages: coupling GIS with wildlife corridor design in conservation planning. Landsc Urban Plan 156:44–58
Selbig WR, Buer N (2018) Hydraulic, water-quality, and temperature performance of three types of permeable pavement under high sediment loading conditions. U.S. Geological Survey. https://doi.org/10.3133/sir20185037
Swingland IR (2013) Biodiversity, definition of. In: Levin SA (ed) Encyclopedia of biodiversity, 2nd edn. Elsevier, Amsterdam, pp 399–410
Zimmer C, Fox B, Dhalla S, Walters M (2012) Low impact development discussion paper. ICF Markbek. https://sustainabletechnologies.ca/app/uploads/2014/09/LID-Discussion-Paper_Nov-2012.pdf. Accessed 18 May 2020
zu Ermgassen SOSE, Utamiputri P, Bennun L, Edwards S, Bull JW (2019) The role of "no net loss" policies in conserving biodiversity threatened by the global infrastructure boom. One Earth 1(3):305–315

Chapter 27
Planting and Landscaping for Sustainability

Abstract Among other benefits, effective planting and landscaping contributes to the cooling of urban areas. They mitigate building practices that often utilize materials which absorb and reflect heat to create extreme temperature fluctuations. These fluctuations can lead to droughts which threaten the biodiversity and resilience. This chapter examines sustainable landscaping design concepts. The chapter first highlights the importance of planting and landscaping for sustainability, then examines strategic planting for weather control, explores the principles of xeriscaping, presents the concept of planting for placemaking, and illustrates the benefits of edible landscapes. It concludes with recommendations for planners of how to incorporate landscaping and planting to ultimately counteract the urban heat island effect and conserve water.

Keywords Effective planting · Weather control · Xeriscaping · Placemaking · Edible landscapes · Urban heat island · Fertilizers · Pesticides · Windbreaks · Shrub height · Windrow of trees · Climate change · Runoff · Shading coverage · Turf area · Water-use zones · Mulch

27.1 The Importance of Planting and Landscaping for Sustainability

In 2019, a pair of heatwaves was responsible for the death of approximately 1500 French citizens (Berlinger 2019). Dramatic changes in temperature and precipitation patterns to climate change have increased the intensity and frequency of heat waves, causing public health issues (Bowler et al. 2010). In response to this threat, one of the viable solutions is to plant and landscape to mitigate the devastating environmental effects of climate change.

Currently, urban landscapes are covered with foreign flora such as grass in private and public spaces (Fig. 27.1). This common practice requires excessive amounts of water, chemicals, and energy that depletes resources and contaminates water. When chemicals—including fertilizers, pesticides, and herbicides—are applied to

A. Friedman, *Fundamentals of Sustainable Urban Design*,
https://doi.org/10.1007/978-3-030-60865-1_27

Fig. 27.1 Public and private landscapes are covered with foreign plant species such as grass like this home in Montreal, Quebec, Canada

turf and non-turf landscapes they pollute groundwater reserves. Runoff also transports inorganic fertilizers into storm sewers where nitrogen and chemicals are leached into water tables, contaminating the surrounding soil and air. Additionally, when fertilizers reach natural streams, they lower dissolved-oxygen levels and release ammonia, which is toxic to fish and wildlife.

One of the challenges of not covering the ground with vegetation at all is the Urban Heat Island (UHI) effect. The UHI refers to the presence of increased temperatures in urban environments compared to the surrounding rural areas. These abnormal temperature differences are caused by ground cover and building materials that absorb and reflect solar radiation (Fig. 27.2). Therefore, it is crucial to design spaces in cities with appropriate materials to control albedo and the input of anthropogenic heat (Bowler et al. 2010).

27.2 Strategic Planting for Weather Control

There are various landscaping solutions to control the prevalence of wind in open spaces. First, planting shrub hedges creates windbreaks. Shrub hedges break approximately 25–60% of wind (Meltzer 2014). Shrubs with this range of permea-

Fig. 27.2 Urban Heat Island is caused by heat absorbing ground cover materials such as asphalt in parking lots and building materials that absorb solar radiation

bility are preferable to solid built barriers as they block wind during colder months and allow moderate wind penetration during the warmer seasons. The ideal shrub height is determined by the size of the area that it is intended to protect; the higher the shrub, the greater the protected area. The typical shrub height of approximately 1 m (4 ft) is optimal because it allows users to comfortably enjoy the protected area without obscuring their view.

To prevent excessive wind from penetrating the barrier, a shrub species with dense, needled hedges instead of a broadleaf species are preferred although native flora should be used. Second, windrows of trees provide wind protection in open spaces. Windrows are single rows of identical trees that are planted closely together. The dense formation of branches and foliage enables windrows of trees to break incoming wind (Fig. 27.3). For the implementation of a windrow of trees, one needs to choose a tree species that is tall and narrow to ensure optimal density and uniformity. Should designers implement tree windrows into a master plan, they need to ensure that windrows, buildings, and roads are aligned according to prevailing wind direction (Meltzer 2014).

Planting and landscaping to control precipitation is essential for water management in open spaces. Climate change causes property damage to urban areas as they become increasingly subject to heavy precipitation within a short period of time. Therefore, when restructuring an open space, it is important to identify sites of rain

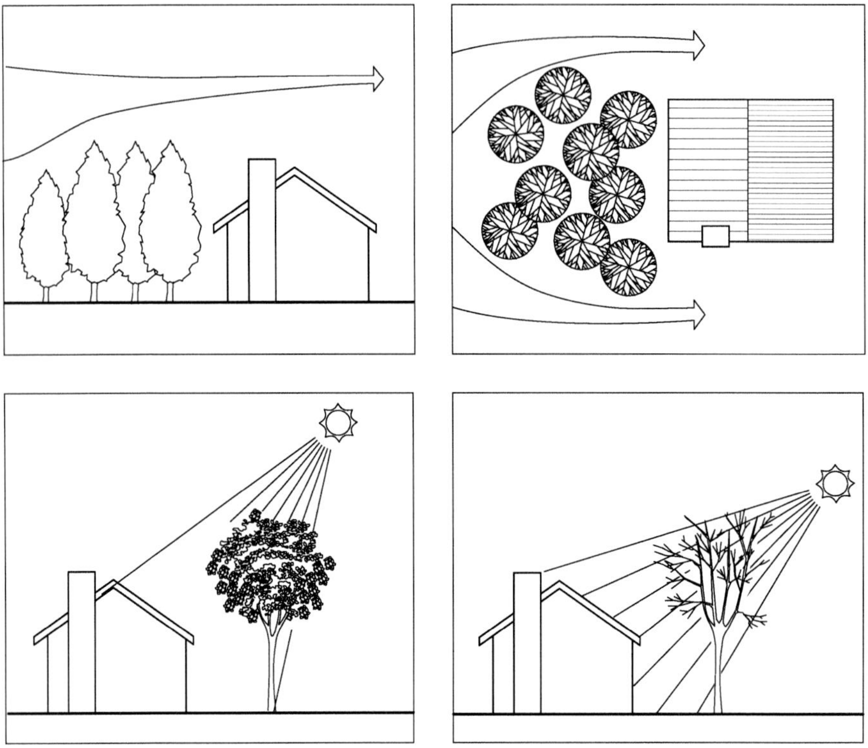

Fig. 27.3 Dense formation and foliage enable windrows of trees to break incoming wind

collection and suitable routes for emergency waterways such as wetlands (Meltzer 2014). As explored in Sect. 6.2, the original topography should be preserved as much as possible and areas with a history of flooding should be studied. Landscape open spaces with materials that absorb precipitations. These absorbent materials collect precipitation from heavy rainfalls, generate groundwater reserves and allow precipitations to percolate and drain into the region's watershed (Fig. 27.4). Implementations that filter precipitation appropriately include dry creek beds and rain gardens. Dry creek beds are trenches that are edged with plants and filled with rocks. They are best used to facilitate drainage and to prevent erosion by reducing runoff. Place the dry creek bed on an existing slope and ensure that water flow leads to a filtration system such as a rain garden rather than toward a street or residential zone (Meltzer 2014). Treat and reuse collected rainwater to avoid the use of freshwater.

Urban greening is a proven strategy to reduce urban air temperatures. The shade generated by trees reduces temperature by blocking solar radiation and preventing the warming of land surface and air (Bowler et al. 2010). To effectively green the urban space, it is important to increase the number of trees and vegetation in parks, plant trees along streets, and increase the amount of living walls on facades. Within the open area, a designer needs to consider major wind barriers such as walls and

Fig. 27.4 When planning open spaces, it is essential to identify sites of rain collection also known as wetlands and suitable routes for emergency waterways as shown here in Montreal, Quebec, Canada

Fig. 27.5 Planting diverse, dynamic, native shrubbery that can withstand hot and dry periods, is recommended to maximize a place's biodiversity as shown here in Porvoo, Finland (left) and Victoria, British Columbia, Canada (right)

fences to increase the amount of cool air that flows through the space. Tree species vary in their ability to reduce air temperature based on their vegetation composition and area of shading coverage. Trees need to be selected based on the amount of solar radiation that they mitigate. Planting diverse, dynamic, native shrubbery that can withstand hot and dry periods, is recommended in order to maximize a place's biodiversity (Fig. 27.5).

Fig. 27.6 Roadside trees be them urban or rural, provide great ecological and aesthetic contributions to urban spaces

Additionally, roadside trees provide valuable ecological and aesthetic contributions to urban spaces (Fig. 27.6). Their planting requires the consideration of six main factors: shape, size and growth, character, robustness, maintenance costs, and trouble-free qualities. For shape, size, and growth, they must: grow upright, have a slender crown and deep roots. The trees' character refers to whether it blends into the roadside scenery effectively and the robustness refers to the tree's strength.

The tree must be able to withstand harsh environmental conditions such as extensive dry periods, frost, and pest attacks as well as remain stable during storms. The consideration of maintenance costs evaluates the tree's life expectancy and the frequency of trimming required. Roadside trees should be planted on the north-east, sunlit side of the street.

27.3 Planting with the Principles of Xeriscaping

A xeriscape is a planned landscape that incorporates sustainable design solutions that reduce or eliminate the need for supplemental water from irrigation, minimize energy consumption and chemical use, and mitigate soil contamination in green spaces (Fig. 27.7). Urban xeriscaping has the potential to reduce water use, tem-

Fig. 27.7 A xeriscape is a planned landscape that reduce or eliminate the need for supplemental water from irrigation, minimize chemical use, and mitigate soil contamination in green spaces

peratures, and outdoor thermal discomfort (Chow and Brazel 2012). The most suitable plants are pest and disease-resistant and adapt to the local climate. The five principles of xeriscaping are suitable plant material, size of turf area, water-use zones, mulch and soil, and irrigation systems and rainwater collection.

In addition, suitable plant material to reduce irrigation need to be chosen. Species must require minimal water and should survive from rainwater alone. A good rule of thumb is to prioritize deep-rooted, native plants as they require less irrigation and survive extended intervals between watering periods. Generally, mature trees and shrubs can withstand water stresses.

The size of the turf area is also an important consideration in xeriscaping; it is vital to minimize the turf area in an open space. Though the absence of turf is optimal, the incorporation of grass is sometimes necessary to cater to specific human activity needs. If turf integration is unavoidable, a sustainable solution is to line turf areas with strong and deep-rooted plants to absorb water.

Third, designers must hydrozone the open space. Hydrozoning is the process of designating water-use zones to ensure efficient xeriscaping (Fig. 27.8). For open spaces, there are three main zones: the oasis, a transition zone, and the low water-use zone. The oasis is the zone with the highest water use. Within the open space users spend the most time in the oasis, as the temperature is cooler. Second, the transition zone receives moderate use. This area contains plants that require less

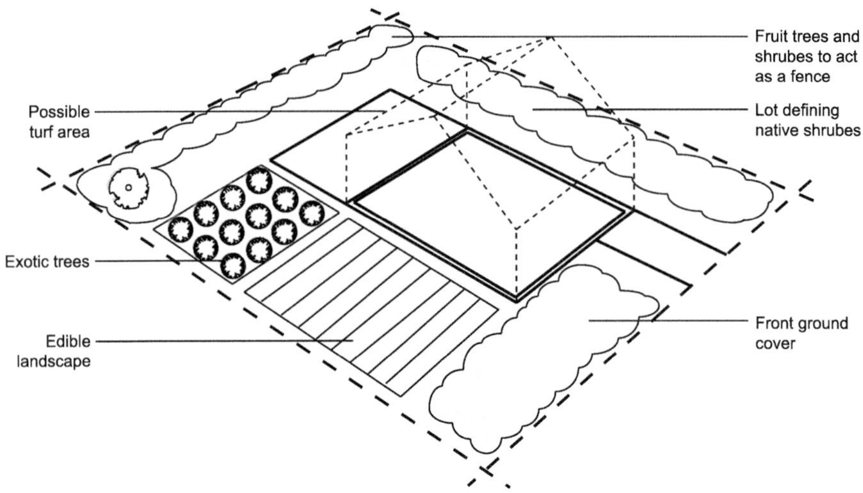

Fig. 27.8 Hydrozoning is the process of designating water-use zones to ensure efficient xeriscaping and water use

frequent irrigation and maintenance. Last, the low water-use zone requires no supplemental water and can survive during prolonged dry periods.

The consideration of mulch and soil is crucial. Mulch is ideal for xeriscaping because it eliminates weed growth, reduces soil compaction and temperature, slows erosion, minimizes evaporation of soil moisture, and decomposes to provide fertile soil (Fig. 27.9). Mulch should be spread out at a maximum depth of 10 cm (4 in.), anything deeper results in oxygen loss. Furthermore, it is essential to amend the soil. Soil amendment is the process of adding elements to soil for the purpose of improving its texture, structure, fertility, porosity, or other qualities. Additionally, soil amendment improves the soil's ability to retain water. In turn it also reduces the amount of water needed to maintain vegetation. Soil amendment should be completed prior to the planting process.

Irrigation and rainwater collection are crucial considerations for xeriscapes. Landscapes will require little to no irrigation if landscapers follow the four principles highlighted above and select appropriate plant species to retain rainwater.

In summary, xeriscaping is a sustainable practice in planting and landscaping for water conservation. Effective xeriscapes balance conservation practices with aesthetic values in the design process.

27.4 Planting for Placemaking

Standardization of the urban design process inhibits the formation of a local identity. Planting and landscaping for placemaking is a viable solution to combat uniformity. Placemaking is the process of creating prosocial environments to give community members a sense of belonging to a place (Fig. 27.10). The practice aims

Fig. 27.9 Mulch eliminates weed growth, reduces soil compaction and temperature, slows erosion, minimizes evaporation of soil moisture, and decomposes to provide fertile soil

Fig. 27.10 Landscaping for placemaking is the process of creating unique environments to instill a sense of belonging to a particular place as shown on the left in Salerno, Italy and Krakow, Poland right

to promote the health and wellbeing of urban dwellers by improving their access to, and experience within, community spaces (Brownett and Evans 2019). Placemaking is important because without cultural or temporal continuity, open spaces lose their ability to sustain a culture.

Landscape architect Dr. Siqing Chen, explores how the implementation of hedgerows in open spaces can establish a sense of place (2009). Designers can implement a hedgerow with an open area network to provide users with multiple ways to access nature. Existing hedgerows are prioritized, and new ones are planted to create a unique arrangement. The repeated use of a specific plant creates a plant network. In this case, hedgerows are implemented throughout the urban fabric to create a network and provide the area with a unique, local identity.

References

Berlinger J (2019) Nearly 1,500 Deaths linked to French heat waves. https://www.cnn.com/2019/09/08/europe/france-heat-wave-deaths-intl-hnk-scli/index.html. Accessed 28 Apr 2020

Bowler DE, Buyung-Ali L, Knight TM, Pullin AS (2010) Urban greening to cool towns and cities: a systematic review of the empirical evidence. Landsc Urban Plan 97(3):147–155

Brownett T, Evans O (2019) Finding common ground: the conception of community arts festivals as spaces for placemaking. Health Place 61:102254

Chen S (2009) Integrating hedgerow into town planning: a framework for sustainable residential development. Int J Hum Social Sci 5(5):293–302

Chow WTL, Brazel AJ (2012) Assessing xeriscaping as a sustainable Heat Island mitigation approach for a Desert City. Build Environ 47:170–181

Meltzer L (2014) Consideration of climate change in the design of parks and open spaces. https://www.hybridparks.eu/wp-content/uploads/downloads/2014/08/Climate_Change_NRW_en.pdf?fbclid=IwAR1ttF471AUzOxe87-LbyX3cNB8lPiSZt3zyOz1YCuKylOWiYMxhP5y6BhM. Accessed 26 Apr 2020

Chapter 28
Open Spaces for Healthy Living

Abstract Rates of obesity and lifestyle-related diseases are steadily increasing worldwide despite ongoing efforts to curb them. These alarming trends are due in part to the current built environment which prioritizes vehicle usage over active mobility by segregating land uses. In response to the increasing global prevalence of inactivity, this chapter explores urban design concepts that promote healthy living. To do so, the chapter explores the obesity and inactivity public health challenge, examines the principles of walkability, suggests solutions for retooling cities accordingly, discusses how to locate and design play areas, makes recommendations for planning to include all ages, and concludes with guidelines for future planners to incorporate into health-conscious urban design.

Keywords Obesity · Lifestyle · Play areas · Walkable communities · Mixed-use developments · Public transit · Public open areas · Green spaces · Wheelchair friendly · Physical abilities · Pocket parks · Mixed-use developments

28.1 Public Health Challenges

The prevalence of obesity in the United States has increased from 30 to 42% in the last 18 years (Hales et al. 2020). This significant increase is partly a result of vehicle-oriented built environments. It has become clear that those who live in lower density settings are more likely drive (Fig. 28.1). As explored in previous chapters, past planning practices force inhabitants to rely on vehicles to accomplish daily activities. According to the American Bureau of Transportation Statistics, this results in approximately 90% of daily trips occurring using private vehicles (2017). Therefore, one strategy to counteract the obesity epidemic, is to prioritize active mobility and health equity by promoting an integrated and inclusive approach to urban planning.

A. Friedman, *Fundamentals of Sustainable Urban Design*,
https://doi.org/10.1007/978-3-030-60865-1_28

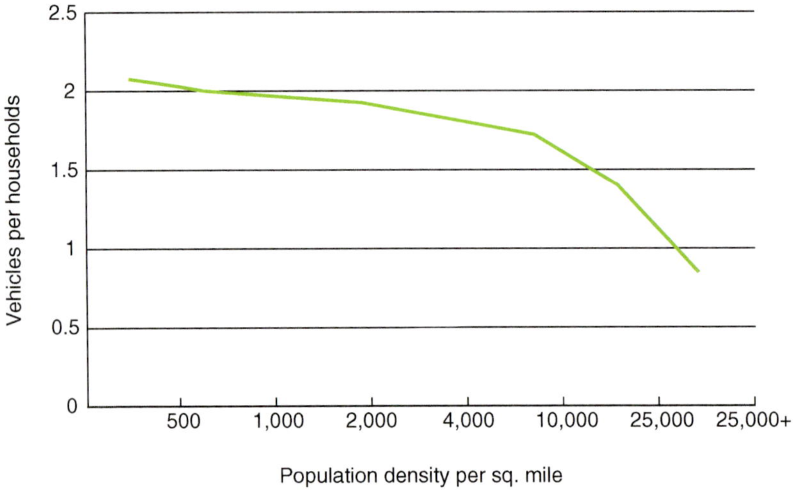

Fig. 28.1 Vehicle ownership per household decreases with increased residential density

28.2 Principles of Planning for Walkability

As noted in previous chapters, creating walkable cities is a vital aspect of encouraging healthy living. Walkable communities embrace a variety of factors that support active mobility including mixed-use land developments, street connectivity, green areas, active transportation, and pedestrian prioritization. It joins a range of policy and design measures that communities can introduce to promote active lifestyle (Fig. 28.2).

As noted in Chap. 9, mixed-use developments encourage active mobility and promote efficient daily habits. Locating services and amenities within walking distance from residences encourages dwellers to walk rather than use vehicles.

Street connectivity is the density of road connections and links in street networks (Mecredy et al. 2011). Communities with highly connected street networks have many intersections, roads with numerous short links as well as few dead ends and cul-de-sacs. High street connectivity in communities signifies the convenience of non-vehicle mobility routes. Ideally, these routes provide pedestrians and cyclists with multiple efficient paths through a city. These paths ought to link urban dwellers to an interconnected network of mixed-use developments, public transit, public open areas, and green spaces.

Having green spaces also support walking. Greenery has an important role in drawing residents outdoors as it creates a comfortable environment for dwellers to leave their homes to walk, play, exercise and engage in leisurely activities (Fig. 28.3). Beyond walkability, research suggests that the presence of trees and grass cover are linked to important social ecosystem indicators including stronger ties among neighbors, a greater sense of safety, more supervision of children in outdoor spaces and fewer crimes (Kuo 2003). Moreover, green, open spaces must be included

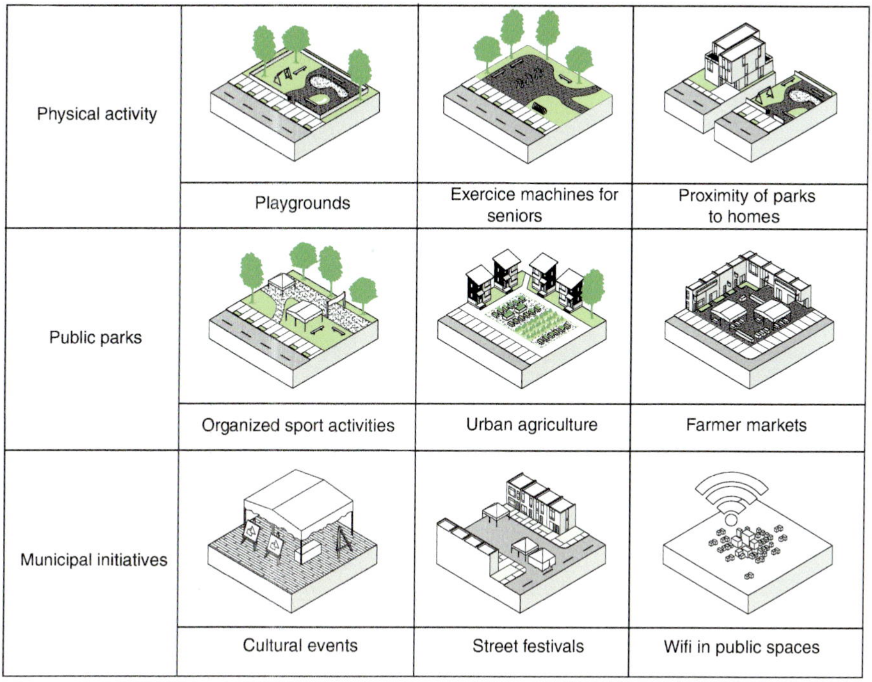

Fig. 28.2 Design and policy measures that communities can introduce to promote active lifestyle

Fig. 28.3 Greenery plays an important role in drawing people outdoors as shown in these Berlin's, Germany images

within an interconnected open space system which allows residents that live on the periphery of a community to reach the other end of the neighborhood with ease. This consideration of green spaces as a system that was discussed in Chap. 24, ensures that residents are in proximity to parks regardless of where they live.

Fig. 28.4 Planning accessible streets will allow people with reduced mobility to spend more time outdoors as shown here in Amsterdam, the Netherlands

Active transportation is the use of human powered mobility methods including walking, biking, and the use of public transport to get around (Tam 2017). Users' ability to engage in active transportation depends on a suitable infrastructure that was put in place to support it. People are more likely to engage in activities outdoors when they feel safe on sidewalks or bike lanes, crossing roads or waiting for and using public transit.

Accessibility for users with physical impairments is crucial. In 2012, for example, 20% of Canadians aged 15 or older reported living with a disability that limited their daily activities (Morris et al. 2018). To address this inequity, urban designers need to plan places that are wheelchair friendly throughout the city. Planning accessible streets and providing rest points will allow disable people to spend more time outdoors (Fig. 28.4). Additionally, creating parks of various sizes near homes will encourage disabled users to be more active.

Finally, having rapid public transit will mitigate the usage of private automobiles. Public transit must be efficient, timely and safe. As outlined in Chap. 20, users should feel comfortable while waiting for public transit that needs to circulate frequently.

28.3 Retooling Outdoor Areas for Healthy Living

Outdoors areas can be thought of as "exercise machines" designed to promote healthy living. This means they should offer active, recreational opportunities for people of all ages and physical abilities. This section explores three solutions for

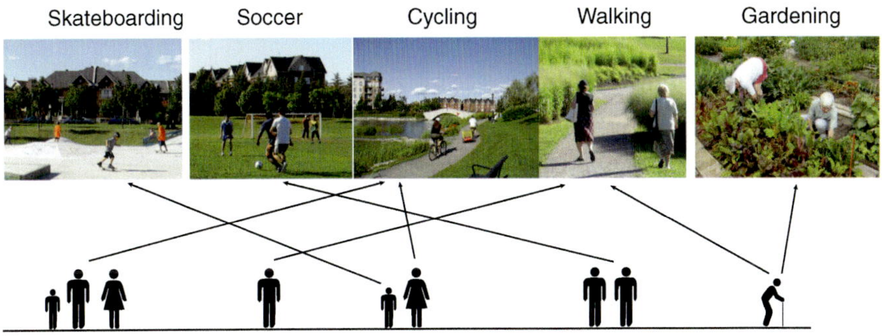

Fig. 28.5 Providing parks of different scales and contents ensures that people have access to activities that meet their specific needs and interests

retooling cities to contribute to healthy living: variety of park sizes, street surveillance, and tools to encourage active transportation.

It is important to plan communities with multiple parks that vary in size as outlined in Chap. 24. Parks need to range from pocket parks to large central community spaces to ensure that urban dwellers have access to those that meet their specific needs and desired activities (Fig. 28.5). Additionally, landscaping implementations must be "breathe easy." This concept means that greenery is allergen-free when possible and appropriate for asthmatic users. Flora with more subtle scents should use used rather than those with very strong odors.

Surveillance in the form of eyes on the street contributes to a safe and engaging environment. Streetlights illuminate spaces at night to increase safety, however, planners should be careful not to implement streetlights in a way that disrupts biodiversity. Build mixed-use developments which feature constant lines of sight to the surrounding public space. These developments should be dense and include 25 units or more per acre (10 units per hectare) with commercial space on the ground level and residences above (Abbasi 2016). This approach ensures the presence of surveilling users at any time of day to make them feel safe, regardless of time (Fig. 28.6).

In Porvoo, Finland, high-quality, rapid bicycle lanes serve as a connection between the city center and the surrounding areas. The high-speed bike lanes are located throughout the city to guarantee that commuters need not cross busy streets. Additionally, planners added solar panels along these bike lanes to mitigate the effects of harsh sun and wind and provide the city with a source of renewable energy. It demonstrates that implementation strategies that encourage healthy living need not be limited to address one function but urban design solutions that can provide a variety of benefits.

Green space also plays an important role in the active lifestyles of residents in Skaftkärr, Finland. Large parks surround the community and small play areas are designated within, offering an inviting environment for children to play and adults to socialize. The provision of green space encourages residents to leave the home—and further, to walk, play sports, and partake in leisurely activities in a comfortable environment. The green space of one park cuts through the middle of the commu-

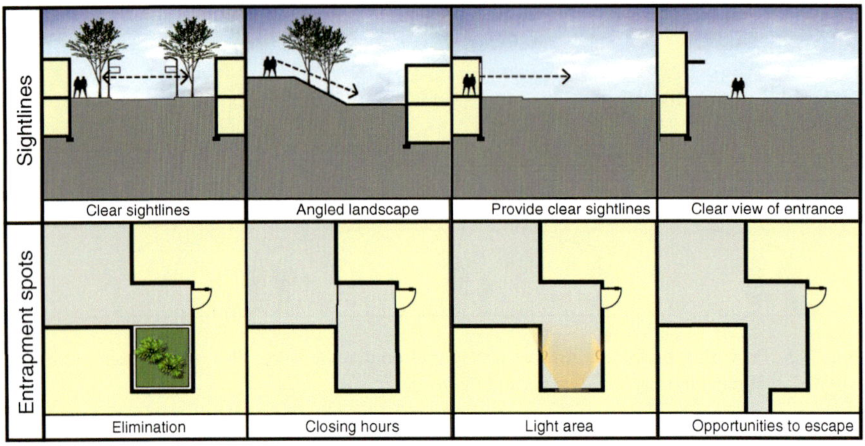

Fig. 28.6 Introducing measures in the form of "eyes on the street" and sightlines and eliminating entrapments spots contributes to a safe and engaging environment

Fig. 28.7 Green public space plays an important role in the active lifestyles of Skaftkärr's residents in Porvoo, Finland

nity, connecting with another park on the opposite end of the neighborhood. This maximizes the accessibility of green space. Residents are always close to a park, whether they are on the edge, or in the center of the community (Fig. 28.7).

28.4 Locating and Designing Play Areas

Playgrounds should be designed and located to promote children's active lifestyles. In the same vein as providing eyes on the street, seeing some engage in physical activity may influence and motivate others to join them. The amount of physical activity children engaged in will depend on the safety, accessibility, and quality of the play area (Fig. 28.8).

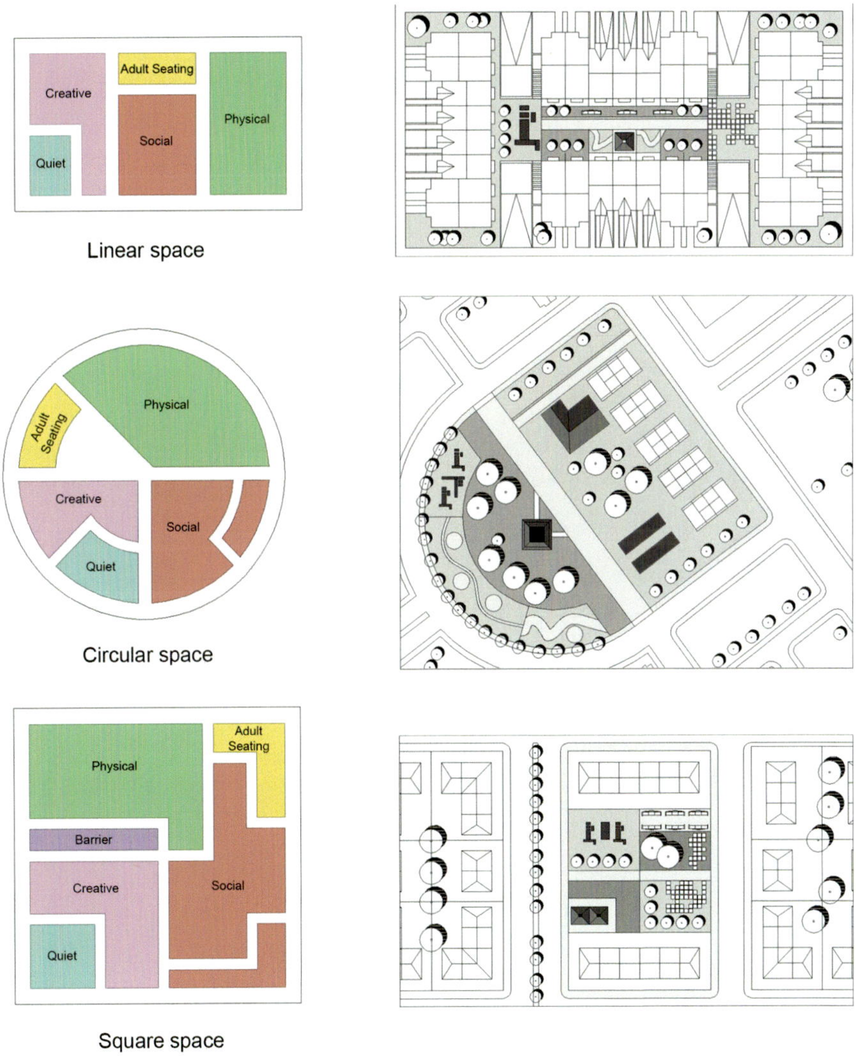

Fig. 28.8 Three conceptual plans of children's play spaces, providing physical, social, creative, and quiet play areas

In the past, children's play areas were often located in residential back lanes and pocket parks. Since then, schools have increasingly been relocated from neighborhood centers to isolated areas that are only accessible by vehicle. Therefore, youth are typically unable to walk or bike to recreational areas. To counteract the physical restrictions caused by displaced playgrounds, design play areas that are weaved into residential, mixed-use zones to ensure that children can access play infrastructure independently and parents are able to supervise them from the comfort of their homes. Additionally, construct infrastructure with sustainable or recycled materials such as wood, cork, and rope.

28.5 Planning for Exercise and Recreation

Space needs to be designated for recreation and exercise in public open spaces. Allocating an open space specifically for physical activity permits a versatile range of actions such as aerobics, flexibility training, dance, and cardio workouts (Fig. 28.9). Overall, exercise in open spaces contributes to general health, improves quality of life, and increases social interaction.

Fig. 28.9 Open space for physical activity permits actions such as aerobics, flexibility training, dance, and much more that contributes to general health, improves quality of life, and increases social interaction as shown in this Singapore image

Fig. 28.10 Seniors participate in exercise in a public park in China

China prioritized senior's wellbeing by increasing public access to exercise activity and equipment. In Beijing for example, since 1998 the city has built more than 4000 outdoor gym facilities in its public parks (Loukaitou-Sideris et al. 2014). Among the ageing population this normalizes public exercise, promotes an active lifestyle, and encourages spending time in outdoor social spaces. The parks' designs feature user-friendly equipment, allowing seniors to exercise comfortably and safely, according to their abilities (Fig. 28.10). Additionally, within many of these parks, designers have implemented chess tables. This facilitates social interaction and encourages mental exercise. Most importantly, this infrastructure is for public use and is free of charge.

Open spaces designed to cater to users of all ages, effectively dismantle social and economic barriers for community engagement and social interaction. This chapter explored how public open spaces can address the increasing prevalence of obesity, social isolation, and extreme weather events within the built environment by promoting healthy living with accessible urban design when implementing the solutions.

References

Abbasi J (2016) As walking movement grows, neighbourhood walkability gains attention. JAMA 316(4):382–383

Bureau of Transportation Statistics (2017) National household travel survey daily travel quick facts. https://www.bts.gov/statistical-products/surveys/national-household-travel-survey-daily-travel-quick-facts. Accessed 19 May 2020

Hales CM, Carroll MD, Fryar CD, Ogden CL (2020) Prevalence of obesity and severe obesity among adults: United States, 2017–2018. NCHS Data Brief. Available via CDC. https://www.cdc.gov/nchs/products/index.htm. Accessed 18 May 2020

Kuo FE (2003) Social aspects of urban forestry: the role of arboriculture in a healthy social ecology. J Arboric 29(3):148–155

Loukaitou-Sideris A, Levy-Storms L, Brozen M (2014) Placemaking for an aging population. https://www.lewis.ucla.edu/wp-content/uploads/sites/2/2015/04/Seniors-and-Parks-8-28-Print_reduced.pdf. Accessed 19 May 2020

Mecredy G, Pickett W, Janssen I (2011) Street connectivity is negatively associated with physical activity in Canadian youth. Int J Environ Res Public Health 8(8):3333–3350

Morris S, Fawcett G, Brisebois L, Hughes J (2018) A demographic, employment and income profile of Canadians with disabilities aged 15 years and over, 2017. https://www150.statcan.gc.ca/n1/pub/89-654-x/89-654-x2018002-eng.htm. Accessed 12 May 2020

Tam T (2017) The chief public health Officer's report on the state of public health in Canada 2017: designing healthy living. Public Health Agency of Canada. https://www.canada.ca/en/public-health/services/publications/chief-public-health-officer-reports-state-public-health-canada/2017-designing-healthy-living.html?wbdisable=true. Accessed 03 May 2020

Chapter 29
Urban Agriculture and Community Gardens

Abstract In response to the increasing global prevalence of food insecurity, unhealthy eating habits, and the fact that food must travel from afar to reach consumers which raises the rates of carbon emissions, this chapter explores design concepts that promote urban agriculture (UA). It first discusses the challenges of the current food system, examines how to create and locate edible landscapes, outlines the best places for farmers' markets and food exchanges, and concludes with guidelines for ways of incorporation into urban design practices.

Keywords Food system · Edible landscapes · Global food system · Urban agriculture · Mental health · Social interactions · Food waste · Greenspaces · Stormwater management · Biodiversity · Rain gardens · Plant species · Community gardens

29.1 The Challenges of Current Urban Food Systems

As a result of the concentration of over half of the world population in urban areas, more food must now travel from rural to urban markets (United Nations 2014). In addition, rates of carbon emissions associated with the energy consumed during food production and distribution are increasing and contributing to climate change (Fig. 29.1). Moreover, the current global food system is showing fragility as it continuously tries to meet constantly increasing demands (Chamie 2020). Furthermore, the current production methods of food negatively affect public health as mass agriculture relies on harmful nitrogen fertilizers to grow produce.

Urban agriculture (UA) is the process of growing edible plants locally. The principles of urban agriculture address and mitigate a variety of concerns related to growing food including packaging, distribution, wholesaling, retailing, restaurants, education systems, local cultures of food, and food security (de la Salle and Holland 2010). Urban agriculture aims to minimize the distance between producer and consumer, encourages consumers to be mindful about what they eat, and promotes shopping locally.

A. Friedman, *Fundamentals of Sustainable Urban Design*,
https://doi.org/10.1007/978-3-030-60865-1_29

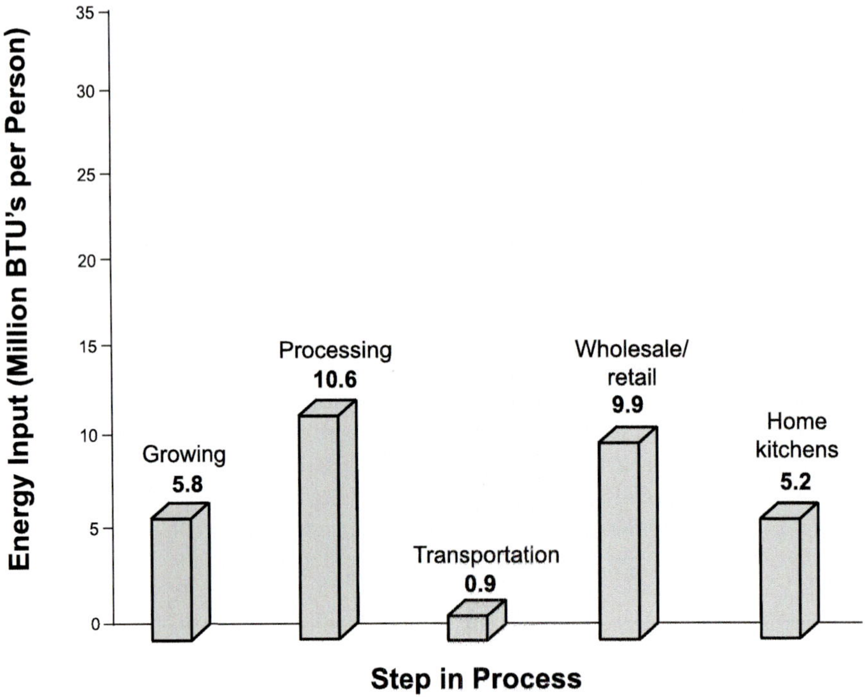

Fig. 29.1 Energy input, measured in BTUs, invested in growing, processing, transporting, and consuming grocery store bought food

Fig. 29.2 Urban agriculture promotes physical well-being by providing a source of engaging exercise for a single person or number of people as shown in these images in Montreal, Quebec, Canada

UA provides a host of benefits for its growers and consumers alike. First, the practice promotes physical well-being by providing a source of engaging exercise for a single person or many (Fig. 29.2). Additionally, participants grow and consume food with high nutritional content. Second, urban agriculture can improve mental health by increasing social interactions, creating a sense of solidarity among neighbors, and providing therapeutic benefits by promoting cognitive, physical, social, emotional, and spiritual well-being (Freymann 2014). Furthermore, the practice has many economic benefits. Urban agriculture creates employment for low-income households, causes consumers to spend less money on groceries, minimizes food waste as crops do not spoil during transit and participants are less likely to discard "ugly produce" if they grow it themselves. Last, urban agriculture is environmentally beneficial because it diversifies urban greenspaces, supports stormwater management, contributes to biodiversity, and mitigates air pollution.

29.2 Creating and Locating Edible Landscapes

Food systems must shift away from the traditional practice of importing food products toward a new perspective that focuses on growing food in urban areas for local consumption. Despite the smaller scale of urban agriculture, growing food in urban areas is efficient because cities are warmer, there are fewer pests and transportation from producer to consumer is minimized. The following section explores how planners can design urban gardens to establish sustainability and maximize efficiency.

There are a variety of factors that planners must consider ensuring that urban gardens are appropriately integrated. First, one needs to examine the surrounding area for prevailing landscaping features. When possible, incorporating existing mature trees into the urban garden will provide necessary shade. Analysis of the site's slope and drainage is also essential (Fig. 29.3). Site drainage should be assessed, noting where rainfall pools occur to identify areas of poor drainage as well as creating rain gardens or plant species that require a lot of water to improve drainage.

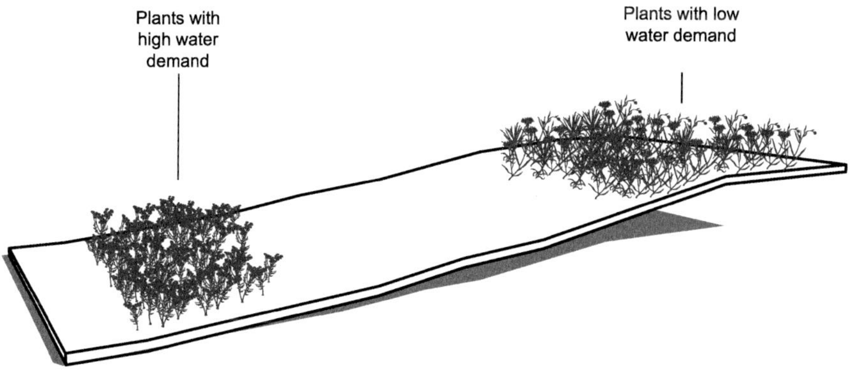

Fig. 29.3 In a sloping terrain, plants that consume little water can be placed atop while those with higher water needs can be planted at the bottom to where the rainwater will flow

Fig. 29.4 In Middlesex Center that was designed by the author, plot for urban agriculture were combined with play and social meeting areas

Planners should locate entrances, exits, pedestrian paths, and choose access points that minimize the distance between the urban garden and streets or other public paths. It is important to consider how users will access the urban garden from transit stops and other community amenities that surround it. Since urban agriculture is an opportunity to encourage social interaction, other amenities should be located near the urban garden, creating a variety of uses for the communal outdoor space. The surrounding areas should include amenities such as covered outdoor spaces, an area for children's play, along with outdoor seating for relaxation and socialization (Fig. 29.4).

Once the urban garden is integrated into the surrounding environment, designers must choose a suitable type of garden for the open space. Community gardens are a common method of implementing edible landscapes. They are single pieces of land located in public, open spaces that are used for collective gardening by a group of people. There are three main types of community gardens: neighborhood, residential, and institutional gardens.

Neighbourhood gardens are places that residents go to grow their own fruits and vegetables. These gardens sit on parcels of private or public land where individual

Fig. 29.5 Urban agriculture can take the form of neighborhood gardens were plots on public land can be assigned (top right), residential gardens at the rear of several homes (bottom left), wooden boxes in single-family homes (top right) and be part of an institution's green roof (bottom right)

plots can be rented by gardeners. Residential gardens are like neighborhood gardens but are shared and maintained collectively by all the residents of an apartment complex. Last, institutional gardens are either publicly or privately owned and provide education, mental or physical rehabilitation services and therapy (Fig. 29.5).

Within the various types of gardens, there are a variety of planting methods. Vertical planting is a viable solution for growers without access to a lot of space. It is advantageous in small spaces as it produces high yield per unit of area. Vertical planting can be done with vine-like plants that are supported by strings on walls or fences. Rooftop gardens or green roofs are ideal for utilizing unused spaces. Prior to construction, one needs to ensure that roofs can support the weight of a garden's soil, plants, and maintenance equipment (Fig. 29.6). Aquaponics is a combination of aquaculture and hydroponics. The practice is a food-production system that rears small aquatic animals such as snails, fish, crayfish, or prawns in tanks. These systems are sustainable as water in the fish tank feeds the plants and the plants clean the water for the fish. Infrastructure for aquaponics can vary in size and can be placed on rooftops.

Along with choosing the appropriate type of community garden, it is equally important to consider the layout of the garden and contemplate how people will walk in it. Creating paths that are specifically designated for high and low traffic and

Fig. 29.6 Green roofs and vertical planting can be considered types of urban agriculture

ensuring that the garden is oriented toward the sun are highly important. To do so, one needs to avoid planting trees north of tall buildings, as those areas are shaded for most of the day but rather orient rows of plants in the north-south direction to maximize sunlight exposure.

29.3 Planning for Farmers' Markets and Food Exchanges

Food exchanges such as farmers' markets are an integral part of urban food production and consumption. They further the goals of urban agriculture by sharing yielded produce with the broader community. Farmers' markets are places where producers meet regularly to sell local produce and goods to consumers. While they are not intended to replace supermarkets, farmers markets are a healthy and sustainable alternative, as they reduce food transportation and allow residents to easily access fresh food while supporting local farmers (Larsen and Gilliland 2009). Additionally, farmers markets and other food exchanges reduce the distance between farmers and consumers. This proximity facilitates a direct relationship between farmer and consumer that translates into a higher level of care during the growing process and greater confidence for the consumer (Fig. 29.7).

Successful farmers' markets should be planned to ensure accessibility for all community members. Currently, accessing farmers markets often requires a private vehicle. There are six factors that are crucial when locating accessible food exchanges: shelter, visibility, safety, accessibility, placemaking, and public transit.

First, having shelter ensures that the users are comfortable regardless of the weather (Fig. 29.8). Locating farmers' markets in open spaces such as parks or squares is preferred. Second, food exchange areas must be visible to a passerby. Locating farmers markets in busy areas where public transportation users, pedestrians, cyclists, and drivers can see the food exchange is recommended. Third, safety must be prioritized by being in an area with high foot traffic with casual surveillance by users. Additionally, the area must include sidewalks, traffic calming features and

Fig. 29.7 Outdoor farmers markets reduce food miles and allow residents to easily access fresh food and support local producers

Fig. 29.8 Locating buildings for indoor farmers markets in busy areas visible by users of public transportation, pedestrians and cyclists is highly recommended

streetlights. Fourth, accessibility for people with reduced mobility is crucial. Ensure that areas of food exchange are wheel-accessible by providing ramps and eliminating stairs. Locate farmers' markets near pedestrian and public transit routes to accommodate all modes of transportation. Fifth, employing placemaking strategies by ensuring that the space is welcoming by provide shade with trees and installing benches is important.

The Riverdale Farmers' Market is in Cabbagetown, a neighborhood in Toronto, Canada. The weekly food exchange takes place in Riverdale Park, a dynamic area with high foot traffic. The area is accessible as the neighborhood's main street sits one block west of public transit stops and there are a variety of bike and pedestrian paths that connect the park with surrounding areas. Additionally, the park is walkable with sidewalks on every surrounding street, along with street calming features such as speed bumps and stop signs. There is adequate surveillance as the farmers' market sits within a residential area. Furthermore, the Riverdale Farmers' Market successfully establishes placemaking in a variety of ways. First, there are many co-located activities to draw users in. Within the park, the food exchange sits adjacent to the Riverdale Farm where urban dwellers can visit animals, a dog park and a playground for children. Second, the entire park is covered by a canopy of trees that provide shade over benches. Third, a block west, there are a variety of local shops to visit.

There are five guidelines that must be adhered to when planning for urban agriculture. First, design strategies that encourage the use of urban agriculture in everyday life within communities' urban fabric must be considered. Second, grass-root urban agricultural initiatives need to be supported by the authorities. Third, citizens need also to support local peri-urban and rural agriculture instead of imported goods. Fourth, marginalized people such as low-income groups, the disabled, and the elderly should be considered at every step of the design process. Finally, the initiators need to strive to implement accessible designs by weaving them into the urban fabric of the community. Put simply, all members of the greater urban area must have the opportunity to engage in and contribute to urban agriculture.

References

Chamie J (2020) World population: 2020 overview. YaleGlobal Online. https://yaleglobal.yale.edu/content/world-population-2020-overview. Accessed 12 Apr 2020

de la Salle J, Holland M (2010) Agricultural urbanism: handbook for building sustainable food systems in 21st century cities. Green Frigate Books, Sheffield

Freymann G (2014) About the CHTA. https://www.chta.ca/about-us.html. Accessed 18 May 2020

Larsen K, Gilliland J (2009) A farmers' market in a Food Desert: evaluating impacts on the price and availability of healthy food. Health Place 15(4):1158–1162

United Nations (2014) World's population increasingly urban with more than half living in urban areas. https://www.un.org/development/desa/en/news/population/world-urbanization-prospects.html. Accessed 5 May 2020

Chapter 30
Urban Design for Social Engagement

Abstract According to recent surveys, two in ten American adults report feeling lonely or socially isolated. One especially poignant survey conducted by the insurance company Cigna found that from a pool of more than 20,000 people, young adults aged 18 to 22 were the loneliest (Rao, Our cities are designed for loneliness, 2018). In response to the increase of social isolation, this chapter explores urban design concepts that promote social interaction in open areas. To do so, the chapter highlights the importance of social interaction and sustainability, discusses the typologies of sociable open spaces and their designs, explores the implementation of performance spaces, examines the potential of urban back lanes to stimulate interactions, and concludes with guidelines for future planners to incorporate into social-interaction-focused urban design.

Keywords Performance spaces · Back lanes · Social isolation · Social capital · Social sustainability · Third places · Outdoor living rooms · Public art · Public art · Sightlines · Placemaking · Graffiti and street art

30.1 Social Interaction and Sustainability

Since the 1950s and the rise of loneliness, or social isolation, has become a public health problem. Studies have found that feeling alone makes our lives shorter, our bodies subject to disease, and our minds more vulnerable to mental illness (Rao 2018). To create cities and neighborhoods that are socially sustainable, it is crucial that planners focus on urban design concepts that are built at the human-scale and promote designs that pay attention to social attributes and foster interaction and connectedness among dwellers (Fig. 30.1).

Throughout history, having a place for social interactions was a key component of planning since these spaces created a sense of community and cohesion. The Athenian Agora is perhaps the most famous as it is widely discussed in literature through the ages (Fig. 30.2). This important area provided the Greeks, and later the Romans, with spaces to debate, arbitrate, and socialize in a central urban spot. There the locals were both the actors and audience constantly judging each other's behav-

A. Friedman, *Fundamentals of Sustainable Urban Design*,
https://doi.org/10.1007/978-3-030-60865-1_30

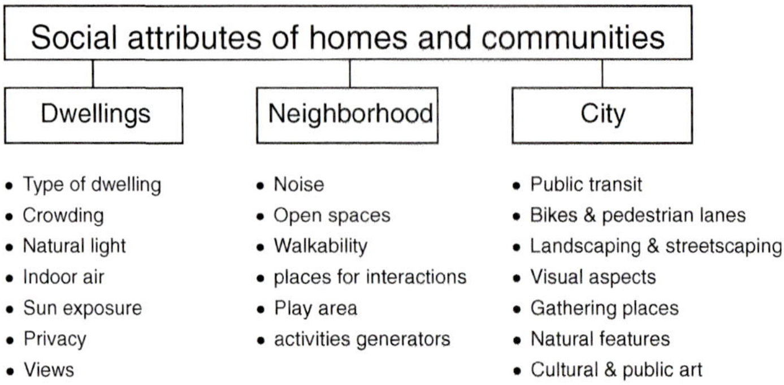

Fig. 30.1 Social attributes of homes and communities

Fig. 30.2 The Athenian Agora Like the one shown in Kamiros, Greece provided the Greeks and later the Romans with spaces to debate, arbitrate, and socialize in an important urban spot

ior. It was also the place where the classes mixed, and citizens could listen to present and future leaders articulate their vision and where social ideas were developed and disseminated. Many historians have described the agora as the ideal public space, and its invention is cited as an important milestone in the creation for urban life and what is known today as social capital.

Social capital refers to the outward-looking open networks that encompass and include our interactions with people from various social groups (Putnam 2000). The concept attempts to quantify how people gather and familiarize themselves with each other. Consistent face-to-face meetings are important because they increase communities' contributions to social capital and promote economic and social sustainability (Putnam 2000).

30.2 Typology of Spaces and Their Design

Small, informal meeting spaces are places that present people with the opportunity to socialize. In these places, there is a social familiarity among their users. Usually, they consist of a diverse group of locals, neighbors, friends, or acquaintances (Simões Aelbrecht 2016).

In his book *The Great Good Place*, Ray Oldenburg refers to such venues, those that are not associated with dwelling or work, as "third places" (1999). They are essential to the secret recipe which turns a cold, sterile neighborhood into a vibrant one. They are not part of an international chain of restaurants, fast food outlets, or sites of exquisite gourmet dining. They are often independently operated, often family owned businesses. Their interior is often worn and shabby yet kept clean by owners who are devoted to the comfort of their patrons.

Visitors are not tied to any schedule in them and they are welcome to come and leave as they please. Third places are also levelers; patrons' wealth, social status or educational background is immaterial and of secondary importance (Fig. 30.3).

Public squares can be considered outdoor living rooms. This is important for creating a strong web of relationships and communal security within an area. Well-designed public squares adhere to the following two rules to foster social interaction. The square can be surrounded on three sides with three-story high, mixed-use buildings, which contributes to placemaking and ensures user comfort (Fig. 30.4). Ideally, commerce is located on the main floor with residential buildings above. Second, the square should have a streetscape to draw users in. For example, squares with a main fountain provide the area with a focal point. Alternatively, include public art to stimulate discussion and gathering. The importance and benefits of public art is discussed in Chap. 31.

Semiformal meeting spaces are spaces where users do not intend to interact but where the built environment is designed in such a way that fosters conversation and are known as "fourth places." Fourth places are different from third places in that they are not grounded in routine but instead accentuate "in-betweenness." Therefore,

Fig. 30.3 Third places are often independently operated—a mom-and-pop type spot shown here near Victoria, British Columbia, Canada (left) and in Daeuville, France (right)

Fig. 30.4 Public squares can be considered as large-scale outdoor community living rooms as shown here in Tel Aviv, Israel (top right), Santiago de Querétaro, Mexico (top right), Bruges, Belgium (bottom left) and Ypres, Belgium (bottom right)

they encompass activities including people-watching, walking, and waiting. Fourth places encourage mundane areas to be more open and foster interaction among a diverse group of users. Additionally, they are socially open because there are no fixed or regular users and the spaces are adaptable to a variety of uses (Simões Aelbrecht 2016). Overall, fourth places are important arenas for social encounters between strangers. These informal fourth places include sidewalks, back lanes, and alleyways (Fig. 30.5).

Among the various types of open public spaces available for socialization, there are several overarching features that they must all abide to. First, the spaces must be safe for users to feel at ease. There should be sightlines from every point, building mixed-use developments and implementing benches for users to rest and socialize. Second, they need to be highly connected with multiple communities to promote walkability. To do so, one needs to ensure that paths are wide to allow the opportunity for the space to be retrofitted (Simões Aelbrecht 2016). Additionally, wide paths allow the completion of activities with minimal conflict and ample space for everyone. Third, there should be amenities such as water fountains throughout open spaces. This makes spaces user-friendly and provides the opportunity for people to stay and linger comfortably.

Fig. 30.5 Fourth places encourage mundane areas to be more open and foster interaction among a diverse set group of users as shown here in Bari, Italy (top left), Mexico City, Mexico (top right), Den Burg, the Netherlands (bottom left) and in Berlin, Germany, (bottom right)

30.3 Performance Spaces

Well-designed performance spaces in open areas will draw users in and foster social interactions. Triangulation is an essential aspect of designing performance spaces; it is the ability for stimuli to attract attention from people who otherwise are disinterested in the social environment (Bagneres 2015). This process stimulates place-making and gives users a reason to attend, enjoy, and return. These urban design implementations for performance enhance community excitement and encourage personal ownership of new public spaces. Performances break preconceived social routines and serve as attractions that become a magnet for people to stop and linger in the space and act as a meaningful catalyst for conversation.

Performance spaces in open areas should adhere to the following five urban design guidelines. First, the performance area must be versatile and able to accommodate a variety of uses. To do so, it is important to build a simple focal point performance stage that clearly communicates where the audience should look and allows a versatility of events and to provide communal seating. It is also necessary to create a tiered seating structure that is multi-purpose and open, using a material that is comfortable for sitting and relaxing. Individualistic seating such as chairs should be avoided to prevent having an audience that is either full or empty.

Additionally, the seating space needs to be designed in a circular formation to facilitate communication and engagement within the crowd. Third, the main performance area, along with the majority of the audience, should be shaded. This ensures that lingering is comfortable during warm months and that the space is usable during colder months. Fourth, one needs to ensure that the main stage and seating is accessible. To do so, it is important to opt for ramps with a gradual incline and handrails. Fifth, prior to designating a performance space, evaluate the topography of the open space. Ideally, seating is placed on a natural slope with the performance stage at its base. Prior to the implementation, evaluate the drainage of the area and address it if necessary (Fig. 30.6).

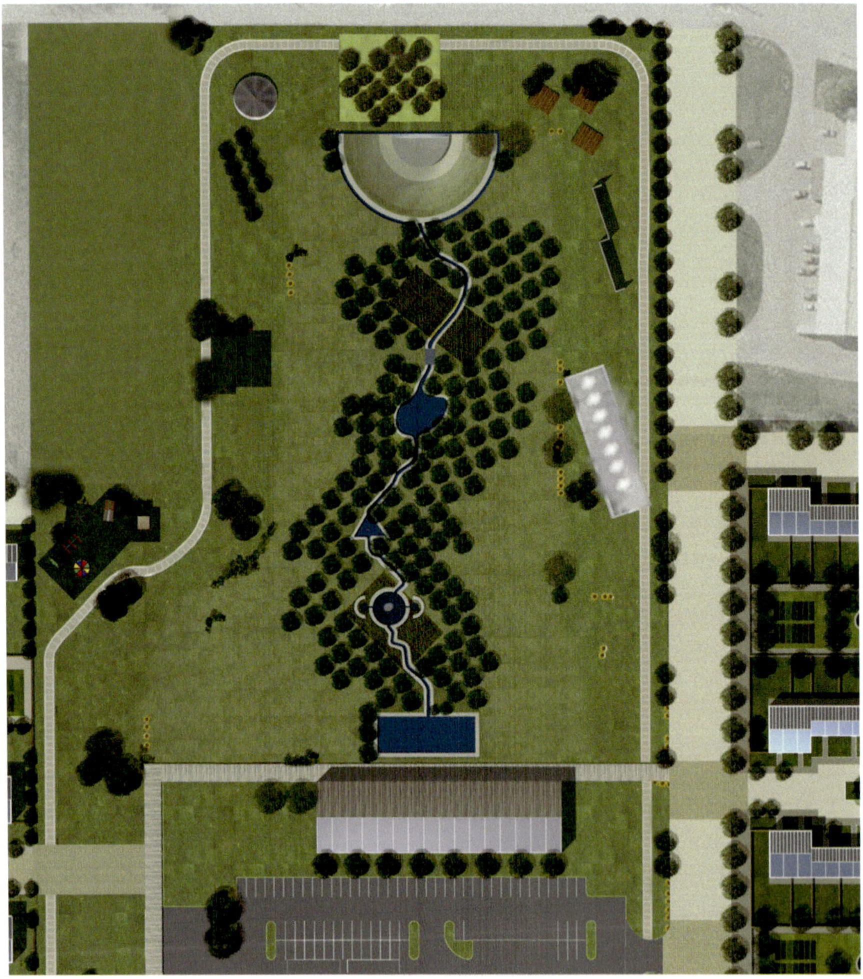

Fig. 30.6 A performance place that was design by the author in Middlesex Center, Ontario, Canada at the end of a water stream

30.4 The Lane and Its Potential in Cities

Back lanes in some cities have an overwhelming potential as social spaces. In North America, existing back lanes are generally used for waste pickup, utility access, resident parking and as an alternative to walking or biking on the street. The potential for back lanes lies in their existing network of paths that connect to peoples' homes. In Europe, Asia, and Oceania back lanes often serve as a refuge from the city and provide the opportunity for pedestrians to walk through a quiet area that is created at the human scale. In this way, they foster a variety of social interaction opportunities such as children playing, growing food or socializing with neighbors (Fig. 30.7).

Back lanes link residential blocks to provide a form of semipublic "commons" (Ismail and Ching 2016). These semipublic places provide an intimate setting for casual social interactions that would otherwise be harder to facilitate in a street-facing front yard. Subsequently, back lanes attract a diverse range of people that can come together to create a lively, sociable, and exciting public environment. The utilization of back lanes is an important reminder for designers to utilize existing infrastructure as opposed to building new.

Fig. 30.7 Lanes can foster a variety of social interaction opportunities such as children playing, growing food or neighbors socializing here in various locations in Montreal, Quebec, Canada

When designing lanes, it is important to ensure that the lane is placed in a strategic location; meaning it is in proximity to several other important gathering nodes. When landscaping, it is also important to take sightlines into consideration to enhance safety. To maximize surveillance, lighting should be installed and trees and bushes should be trimmed to avoid overhang into the back lane. Third, the place should be comfortable for pedestrians; planting tall trees provides shade from high above, installing benches, providing green infrastructure to cool the space, and paving to cover potholes all improve user comfort. Last, engaging local artists to provide aesthetic designs to draw people in with graffiti and street art.

The back lanes in Melbourne, Australia are example of good placemaking. Throughout Melbourne, there are over 40 laneways where previously unused spaces are transformed into commercial enterprises including small bars, clubs, cafes, restaurants, and specialty shops. Today, back lanes are the center of attraction for Melbourne's night-time entertainment and socialization. Center Place is surrounded by five story high buildings and many places to sit. Along with the cafes and boutiques, users also return to enjoy the ever-changing stencil art and graffiti (Fig. 30.8).

This chapter explored how the implementation of urban design features that promote social interaction within cities can counteract the "epidemic of loneliness." When implementing the aforementioned solutions, there are two guidelines that must be adhered to. First, plan implementations to make engaging in social spaces the convenient choice. To do so, ensure that mixed-use development is the predominant residential style in cities and that routes from residential spaces to the heart of the city is a sociable, pedestrian path. Second, ensure that open spaces are accessible for all members of the community, specifically marginalized groups that are often forgotten in planning practices.

Fig. 30.8 The back lanes in Melbourne, Australia are example of good placemaking. Throughout the city, there are over 40 laneways where previously unused spaces are transformed into commercial enterprises including small bars, clubs, cafes, restaurants, and specialty shops and places for gratify artist

References

Bagneres L (2015) Triangulation & public space: bringing citizens together. Ecourbanism Research Network. https://ecourbanismresearchnetwork.com/2015/08/18/triangulation-public-space-bringing-citizens-together/. Accessed 18 May 2020

Ismail WHW, Ching LH (2016) Back lanes as social spaces in Chinatown, Kuala Lumpur. Environ Behav Proc J 1(3):293–299

Oldenburg R (1999) The great good place: cafes, coffee shops, bookstores, bars, hair salons, and other hangouts at the heart of a community. Da Capo Press, Cambridge

Putnam RD (2000) Bowling alone: the collapse and revival of American community. Simon & Schuster, New York

Rao A (2018) Our cities are designed for loneliness. https://www.vice.com/en_ca/article/kzvzpv/our-cities-are-designed-for-loneliness-v25n4. Accessed 18 May 2020

Simões Aelbrecht P (2016) 'Fourth places': the contemporary public settings for informal social interaction among strangers. J Urban Des 21(1):124–152

Chapter 31
Public Art and Street Furniture

Abstract The global economy's gradual shift to the service sector has caused cities to prioritize creativity in their urban design to stimulate growth. Creativity in this context refers to Richard Florida's concept of the "3 T's," technology, tolerance, and talent (Bloomberg, The tree T's of growth, 2001). Cities can achieve creativity by attracting individuals that are technologically savvy or professionally talented and by encouraging a tolerant city culture that embraces heterogeneous demographics. In doing so, the implementation of avant-garde public art and street furnishings plays a fundamental role in creating attractive and desirable cities. In response to this increasing demand for creative urban design, this chapter explores concepts that boost cities' creative index. The chapter highlights the importance of promoting local culture, explores the urban design concepts necessary for successful public art installations, discusses the importance of street furnishings and their considerations, and concludes with guidelines for their incorporation into forward thinking, creative-focused urban design.

Keywords Street furnishings · Creative-focused urban design · Public art · Culture · Play and creativity · Installing artwork · Involvement · Interactive nature · Roadside utilization · Artistic expression · Cohesive space

31.1 The Importance of Promoting Local Culture

The increasing prevalence of globalization puts pressure on major world and old-industrial cities to create attractive and alluring places, and new niches in the urban market to attract residents, visitors, and investors (Akkar 2003). In response to this challenge, the implementation of public art in open spaces is a viable way to create a more welcoming city. In addition to creating vibrancy in existing open spaces, public art promotes local culture and functions as an important city-building tool. Ultimately, public art not only enhances globalization but also pays homage to a places' local culture, defines communities, and contributes to a city's growth and dynamism (Fig. 31.1).

A. Friedman, *Fundamentals of Sustainable Urban Design*,
https://doi.org/10.1007/978-3-030-60865-1_31

Fig. 31.1 Public art pays homage to places' local culture, defines communities, and contributes to a city's growth and dynamism here in Montreal, Quebec, Canada (left) and Brussels, Belgium (right)

Fig. 31.2 Public art can enhance the spiritual and cultural value of a space and serve as an important source of free education as shown in this bench in Vancouver, British Columbia, Canada

Along with the ability of public art to promote local culture, its implementation provides a host of benefits for the community. Public art can promote empowerment, reduce crime and vandalism, increase social interactions, stimulate play and creativity, enhance the spiritual and cultural value of a space, and serve as an important source of public education (Fig. 31.2). The connective nature of public art ignores personal prejudices and serves as a medium of community-building. Unlike private art which entails an intentional, intimate exchange, public art is necessarily presented publicly in diverse social settings to a variety of viewers (Hein 2006).

Fig. 31.3 In Iqaluit, Nunavut, in the Canadian arctic viewing public art is a cultural experience. Along the city's main street lie several life-sized soapstone sculptures depicting wild animals native to the area, all made of soapstone, a local material sourced nearby

Public art's ability to highlight local culture is particularly important as a form of historical and artistic expression for marginalized populations. Iqaluit, in Nunavut, Canada, has made viewing public art a city-wide cultural experience. Along the city's main street lie several life-sized soapstone sculptures—a locally sourced material—depicting aspects of indigenous culture such as wild animals native to the area (Fig. 31.3).

31.2 Urban Design for Public Art

Much of public artworks' successful reception relies on proper installation within the surrounding environment. When installing artwork in an open space, there are three overarching guidelines. First, to ensure safety, art should be placed away from streets, specifically where crowds can gather. This encourages pedestrianism and invites people to linger in the open spaces (Fig. 31.4). Otherwise, groups of viewers run the risk of blocking drivers' views or being unsafely close to traffic. Second, art

Fig. 31.4 Public art can encourage pedestrianism and invites people to linger in the open spaces as shown here in Singapore (left) and Melbourne, Australia (right)

installations must integrate other existing elements of the streetscape. This includes streetlamps, benches, garbage bins, and utility boxes (Yücel 2013). Third, installations must be of good quality. Artworks must have the appropriate surface finish to ensure that pieces are designed for minimal maintenance and can resist vandalism.

Painting intersections and crosswalks is an easy way to create vibrancy in open spaces. Research suggests that this artistic implementation slows down traffic, encourages walking and biking, and increases neighborhood involvement (Nelson 2011). The implementation of this artistic artifact can involve local artists and residents alike.

Interactive public art installations should successfully engage a passerby. Attention catching public installations do the important work of engaging users to facilitate social interaction. An example of interactive public art is the love lock bridges in many cities. These bridges are a popular destination which attracts users to bring personal locks to attach to the railings (Fig. 31.5). Another example of interactive public art is community kiosks. Community kiosks are small pavilions located in public areas that have vibrant patterns and captivating designs. Cork boards are attached to allow the public to post community-related events and news. The interactive nature of their designs allows for the dissemination of information and enables public interactions.

Murals are popular public art installations that require minimal materials and have the dramatic ability to revamp entire spaces. In this case, buildings' blank walls serve as empty canvases for local artists to create. The results have lasting effects on the neighborhood and provide important colorful flare throughout the year.

A famous example of the power of commemorating a person's legacy is in Warsaw, Poland. Poland lost its independence in 1918 and turned to history, music, and religion as a nation-building strategy. In doing so, the Polish composer Frédéric Chopin emerged as a symbol of their national struggle. To this day, Chopin remains a fundamental figure in the Polish nationalist movement and is symbolic of their unique cultural music style (Fig. 31.6) (Cintron 2014). To commemorate Chopin's legacy, benches that plays his music were installed throughout Warsaw. These public art implementations continue to celebrate Chopin's work and his fundamental contributions to Polish independence and culture.

Fig. 31.5 An example of interactive public art is the love lock bridges found in many cities to which people attach their locks as shown here in Krakow, Poland

Fig. 31.6 A public bench in Warsaw, Poland which plays Chopin's music

31.3 The Importance of Street Furnishing

Street furnishings (also known as street furniture) are infrastructure placed on streets that contribute to roadside utilization and to the better usage of the streets, creating living spaces that improve the quality of life for users (Radwan and Morsy 2017). Examples of street furnishings include benches, drinking fountains, garbage cans, and light posts (Fig. 31.7). Street furnishings create comfortable environments

Fig. 31.7 Same style street furnishing in Montreal, Quebec, Canada

where people are encouraged to rest, sit, and socialize with others. The proper selection and implementation of street furnishings creates a spatial identity to establish placemaking, draw users outdoors, and communicates to pedestrians that they are welcome to linger in the space and engage in leisurely activity (Radwan and Morsy 2017). Overall, their successful implementation creates convenient, publicly accessible, and easily maintainable amenities that do not disrupt pedestrian or traffic flow (Yücel 2013).

When thinking about public art, we must consider that its implementation in the urban context occurs on two different levels. On the microscale, each artistic installation functions in the traditional sense. However, on a larger scale, the city itself has an artistic expression. This means that the everyday street furnishings that makeup the built environment of cities including balconies, doorways, iron work on windows, and rooftop edges all meaningfully serve as aspects of the city's artwork (Fig. 31.8).

Prior to the implementation of street furnishings, there are considerations that designers must contemplate. After defining the type of space that is under consideration (e.g., park, street, plaza, waterfront), designers need then to decide who the users of the space will be and what function it will serve. Last, designers can give thought to what times of day the space will primarily be used. Overall, successful street furnishings must be implemented within a greater design concept to create a cohesive space.

Fig. 31.8 The built environment of cities including balconies, doorways, iron work on windows, and rooftop edges all serve as aspects of the city's artwork as shown here in Montreal, Quebec, Canada

31.4 Important Considerations for Street Furnishings

To establish cohesion with furnishings, public art, and existing open space design, there are several important considerations to keep in mind before implementing street furnishings.

First, the placement of furniture is as important as the design itself. To optimize functionality, furnishings can be located based on their intended usage so that they are coherent with existing built infrastructure. Additionally, the furniture needs to be placed in areas with high pedestrianism, especially in places where walkers may wait. All furniture implementations should be placed with the intention of enabling triangulation (as discussed in Sect. 30.3), as an organic way of bringing people together. As a general guideline, street furnishings should be placed at least: 0.45 m (1.5 ft) from curbs' edges, 0.61 m (2 ft) from any garage or wheelchair ramp, 1.2 m (4 ft) from a ramp landing, and 1.5 m (5 ft) from a fire hydrant (Better Streets n.d.).

It is imperative that street furnishings are constructed with materials designed to prevent injury. Designers must ensure that street furnishings are structurally sound. Meaning, they are embedded into the ground or are secured with anchor belts and have no sharp edges or exposed fasteners. When choosing materials, designers must consider the area's climate and weather effects including sunlight (which causes expansion and contraction), wind stress, moisture, salt spray, and ice. Popular materials include steel, wood, stone, concrete and recycled plastic (Fig. 31.9). Ideally designers should implement photocatalytic concrete as a sustainable material. Upon its implementation, photocatalytic concrete reduces air pollution, self-cleans and

Fig. 31.9 When choosing materials, designers must consider the area's weather effects on popular materials such as stone (in Tel Aviv, Israel, left) and steel (in Melbourne, Australia, right)

disinfects itself (Han et al. 2017). The usage of this type of concrete makes for cleaner spaces and allows for the implementation of street furniture that does not require extensive maintenance. All the materials mentioned above differ in their composition, ability to handle weather conditions, and levels of maintenance. Therefore, designers must consider these factors before selecting a material.

Beyond photocatalytic concrete, selecting sustainable materials is important. Designers should select sustainable materials that possess one or more of the following three qualities. First, designers can use materials composed of reused content. Second, designers can use materials that are regionally sourced. These include materials or items that have been gathered, separated, or produced within 805 km (500 miles) of the location of implementation (Radwan and Morsy 2017). Third, planners can utilize rapidly renewable materials including those which are created with plants that are gathered within a 10-year life cycle, such as bamboo (Radwan and Morsy 2017). The proper inclusion of these sustainable substances conserves natural resources and reduces carbon footprints.

After deciding on placement and sourcing sustainable materials, designers must shift their focus to creating visually captivating street furnishings. When considering the visual appeal of street furnishings, color is perhaps the most essential component. Designers should ensure that street furniture's' colors greatly contrast against their background. Additionally, Yücel suggests that the chosen color should have a minimum luminance contrast of 30% to increase furnishings' visibility to passersby's (2013).

Texture is an important aesthetic and functional consideration in the design process of street furnishings. The assortment of materials used to create texture make for fascinating surfaces and conceal small aesthetic faults. Functionally, the implementation of texture benefits children and users with visual impairments. For example, texturized tree grates protect trees to enhance growth and alert users with visual impairments that they are approaching a zone with an obstacle.

Beyond appearance, it is also vital that street furnishing implementations accommodate users with accessibility restrictions. To do so, seating and areas for relaxation must be located near public toilets and telephones. To accommodate users in wheelchairs, tables must be 0.75 to 0.90 m (2.4 to 3 ft) in height with a minimum height of 0.6 m (2 ft) under the table (Better Streets n.d.). Additionally, rest areas that include benches should allocate at least 1 m (3.3 ft) of adjacent space for wheelchair users (Better Streets n.d.).

31.5 Exemplary Street Furnishing Implementations

Seating is important for individuals to see and be seen (Radwan and Morsy 2017). For efficient seating design, place seating furniture in areas where waiting, meeting, or socializing feels natural. Seating must be coherent with the rest of the area and be designed in such a way that users do not feel isolated when other seats are not in use.

For the comfortability of all users, provide armrests and back support. This encourages users to linger for longer periods of time.

Bike racks that are functional, secure, and appealing can promote cycling. Designers must create secure racks where users can lock the bicycle's frame and wheels with a bike lock. To ensure sustainability, functionality, and innovative design, designers should utilize material such as stainless steel, reused plastic, or thermoplastic. Choose a material based on the rack's location, desired aesthetic and the area's weather conditions.

Drinking fountains that are in open spaces can attract users and invite them to linger (Fig. 31.10). This is particularly important in cities that have hot weather. Drinking fountains must be placed in areas with high pedestrian density. Ensure that they are an accessible height for children and users in wheelchairs. Build water fountains with stone, concrete, brick, or metal (copper, bronze, cast iron or steel). Last, ensure that the ground below allows good drainage to avoid slippery surfaces.

Bollards are vertical barriers that delineate open spaces usages. Typically, they are implemented to prevent vehicle access to a pedestrian area or to make streets narrower in efforts of reducing vehicle speed. Designers may include lights on bollards to improve visibility at night and create a safer environment. Designers can beautify bollards to make them appealing yet functional.

Parklets are walkable expansions that give pedestrians more space for leisurely activity in open spaces (Fig. 31.11). Parklets are typically implemented as sidewalk off shoots that extend onto curb-side parking spots. They are constructed at the level of the existing walkways to create a continuous space. Parklets offer pedestrians a

Fig. 31.10 Drinking fountains must be placed in areas with high pedestrian density and ensure that they have an accessible height for children and users in wheelchairs such as these in Vienna, Austria

Fig. 31.11 Parklet is a walkable expansion that gives pedestrians more space for leisurely activity like this one in Montreal, Quebec, Canada

spot to relax, sit, and rest. The space can be filled with seating, greenery, and artwork.

The King Street Causeway in Toronto, Canada provides a notable instance of a successful parklet implementation. The King Street Causeway is a vibrant parklet located at the heart of Toronto's popular downtown strip, King West. The parklet was implemented soon after the launch of King Street's pilot project. The pilot project aimed to make King Street more pedestrian friendly. Since its implementation, private vehicles are now limited to driving up to one block without turning, which prioritizes and encourages the use of public transportation, biking, and walking. The implementation of the pilot project has made King Street more pedestrian friendly and created a calmer environment. Upon the implementation of the pilot project, the City of Toronto created a contest for local urban designers to create a parklet, of which the King Street Causeway was awarded to the winner.

In response to cities' desire to boost their creative index to attract investment, this chapter explored urban design considerations for the installation of public art and street furnishings. When implementing these solutions, there are three guidelines that designers must adhere to. First, designers must remember the critical role of public art as a medium for the inclusion and empowerment of communities. It is important for designers to observe a spaces' existing public art and consider the significance that they hold for the community at large. In doing so, designers should explore the ways that future art implementations can coexist alongside present ones to function as mechanisms for marginalized communities to reclaim public space. Second, designers must remember that the addition of street furnishings in an open space need not follow a top-down implementation. Instead of a single central designer, planners should always consider how to collaborate with the public to select designs and implementations. If locals are involved in the planning process, their communities are more likely to positively receive new installations and enjoy the open spaces they reside in. In this way, community involvement encourages pedestrianism, as users are more willing to enjoy art that they had input in. Last, street furnishings must always be designed and implemented with the needs of the disabled, elderly and children in mind, to ensure the equitable accessibility and enjoyment of urban design for all.

References

Akkar ZM (2003) Regenerating and marketing cities and their 'Public' spaces: a British way. In: Public art & Urban Design: interdisciplinary and social perspectives, pp 22–36

Better Streets (n.d.) Street furniture overview. https://www.sfbetterstreets.org/find-project-types/streetscape-elements/street-furniture-overview/. Accessed 20 May 2020

Cintron PA (2014) Chopin mazurkas and its influence on polish nationalism. Dissertation, Liberty University

Han B, Zhang L, Ou J (2017) Photocatalytic concrete. In: Han B, Zhang L, Ou J (eds) Smart and multifunctional concrete toward sustainable infrastructures. Springer, Singapore, pp 299–311

Hein HS (2006) Public art: thinking museums differently. Rowman Altamira Press, Lanham

Nelson A (2011) Coloring inside the lanes: art that creates community. https://grist.org/cities/2011-12-02-coloring-inside-the-lanes-art-community/. Accessed 15 May 2020

Radwan AH, Morsy AAG (2017) The importance of integrating street furniture in the visual image of the city. Int J Mod Eng Res 9:2

Yücel GF (2013) Street furniture and amenities: designing the user-oriented urban landscape. In: Özyavuz M (ed) Advances in landscape architecture. Namık Kemal University, Turkey

Illustration Credits

Figures not listed below are in the public domain or have been conceived, drawn, or photographed by the author and/or members of his research and design teams. Their names are listed in the Acknowledgments and the Projects' Teams pages. Every effort has been made to list all contributors and sources. In case of omission, the author and the publisher will include appropriate acknowledgment or correction in any subsequent edition of this book.

Fig. 1.4 After Statistics Canada (2006) CYB Overview 2006: Population and Demography

Fig. 1.5 After Alter, L. (2009) "Graph of the Day: Driving vs. Residential Density," www.treehugger.com/files/2009/04/driving-vs-density.php Accessed May, 2016

Fig. 13.2 Courtesy of David Vandervort Architects

Fig. 16.2 After Hodges, T. (2010). Public Transportations Role in Responding to Climate Change. The Federal Transit Administration, U.S. Department of Transportation

Fig. 19.5 Courtesy of Hackland and Dore Architects

Fig. 21.1 Courtesy of Davis Studio Architecture and Design

Fig. 26.7 Courtesy of Wicks Architecture and Design

Fig. 29.1 After Kourik, R. 2004. Designing and Maintaining your Edible Landscape Naturally. Hampshire, U.K.: Permanent Publications

A. Friedman, *Fundamentals of Sustainable Urban Design*,
https://doi.org/10.1007/978-3-030-60865-1

Projects' Teams

I would like to thank those who contributed to the design of the projects cited in the book. I have attempted to recall them all. If I have omitted someone, my sincere apology. I will do my best to correct it in future editions.

Stony Plain, Alberta
Avi Friedman, Architect
Juan Mesa
Fa Xivong Wu

Fort Saskatchewan
Na Zhang
Dutch Osbourne
Jaime Gomez
Jing Yan Liu
Jeff Jerome
Avi Friedman and Louis Pretty, Advisors/Project Leaders

Ponoka, Alberta
Avi Friedman, Architect
Fa Xivong Wu

Middlesex Centre, Ontario
Avi Friedman, Architect
Josie White
Nyd Garavito-Bruhn

Lethbridge, Alberta
Rui Li
Hui Peng
Cherri Pitre
Yanan Li
Yan Li
Ying Le

A. Friedman, *Fundamentals of Sustainable Urban Design*,
https://doi.org/10.1007/978-3-030-60865-1

Xuekun Feng
Xiaoyang Zhou
Kunfei Chen
Avi Friedman and Louis Pretty, Advisors/Project Leaders
Langford, British Columbia
Avi Friedman, Architect
Isabella Rubial
Giacomo Valzania
Peace River, Alberta
Avi Friedman, Architect
Renier Silva
Cynthia Nei

Bibliography

Abbasi J (2016) As walking movement grows, neighbourhood walkability gains attention. JAMA 316(4):382–383
Adams J, Greenwood D, Thomashow M, Russ A (2016) Sense of place. https://www.thenatureofcities.com/2016/05/26/sense-of-place/. Accessed 22 Jan 2020
Agence France-Presse (2017) Dutch electric trains become 100% powered by wind energy. https://www.theguardian.com/world/2017/jan/10/dutch-trains-100-percent-wind-powered-ns. Accessed 08 Apr 2020
Ahmed MM, Sulaiman M (2003) Design and proper sizing of solar energy schemes for electricity production in Malaysia. In: Proceedings. National Power Engineering Conference, pp 268–271
Akkar ZM (2003) Regenerating and marketing cities and their 'Public' spaces: a British way. In: Public art & urban design: interdisciplinary and social perspectives, pp 22–36
Al Nuaimi E, Al Neyadi H, Mohamed N, Al-Jaroodi J (2015) Applications of big data to smart cities. J Internet Serv Appl 6(1):25
American Public Transportation Association (APTA) (2016) Shared mobility and the transformation of public transit. https://www.apta.com/wp-content/uploads/Resources/resources/reportsandpublications/Documents/APTA-Shared-Mobility.pdf. Accessed 08 Apr 2020
ANS Construction LLC (2020) Why should you choose Asphalt instead of concrete? https://ansconstructionllc.com/choose-asphalt-instead-concrete/. Accessed 11 Feb 2020
Arup, Meridian Planning, Planning Alliance (2010) Port Whitby Sustainable Community Plan. https://www.whitby.ca/en/townhall/portwhitbysustainablecommunityplan.asp. Accessed 06 Feb 2020
Averly J, Eley J, Stewart P, Nicolaou L (2004) Creating successful masterplans. Commission for Architecture and the Built Environment, London
Bagneres L (2015) Triangulation & public space: bringing citizens together. Ecourbanism Research Network. https://ecourbanismresearchnetwork.com/2015/08/18/triangulation-public-space-bringing-citizens-together/. Accessed 18 May 2020
Bartoli A, Hernandez-Serrano J, Soriano M, Dohler M, Kountouris A, Barthel D (2011) Security and privacy in your smart city. Paper presented at proceedings of Barcelona smart cities congress, Barcelona, Spain. Accessed 29 Dec 2011
Barton J, Rogerson M (2017) The importance of greenspace for mental health. BJPsych Int 14(4):79–81
BC Transit (2018) BC transit infrastructure design summary. https://www.bctransit.com/documents/1507213895398. Accessed 23 Mar 2020

A. Friedman, *Fundamentals of Sustainable Urban Design*,
https://doi.org/10.1007/978-3-030-60865-1

Belzer D, Autler G (2002) Transit oriented development: moving from rhetoric to reality. Brookings Institution Center on Urban and Metropolitan Policy, Washington, pp 6–15

Ben-Joseph E (1995) Changing the residential street scene: adapting the shared street (Woonerf) concept to the suburban environment. J Am Plann Assoc 61(4):504–515

Ben-Joseph E, Gordon D (2000) Hexagonal planning in theory and practice. J Urban Des 5(3):237–265

Berlinger J (2019) Nearly 1,500 deaths linked to French heat waves. https://www.cnn.com/2019/09/08/europe/france-heat-wave-deaths-intl-hnk-scli/index.html. Accessed 28 Apr 2020

Bertaud A (2004) The spatial organization of cities: deliberate outcome or unforeseen consequence? https://escholarship.org/uc/item/5vb4w9wb. Accessed 01 Feb 2020

Better Streets (n.d.) Street furniture overview. https://www.sfbetterstreets.org/find-project-types/streetscape-elements/street-furniture-overview/. Accessed 20 May 2020

Bilgili BC, Gökyer E (2012) Urban green space system planning, landscape planning. Ecol Appl 22(1):349–360

Bindu CA, Mohamed AR (2016) Water bodies as a catalyst to growth and development-the Case of Kodungallur town, Kerala. Proc Technol 24:1790–1800

Bliss L (2019) The hot new things in dockless electric scooters: docks. https://www.citylab.com/transportation/2019/03/electric-scooters-parking-charging-docks-lime-bird-lyft-spin/584332/. Accessed 08 Apr 2020

Bloomberg (2001) The tree T's of growth. https://www.bloomberg.com/news/articles/2001-04-19/the-three-ts-of-growth. Accessed 20 May 2020

Boname P, Zhu J, Matheson C (2009) Efficient cities: the interrelationship between effective rapid transit systems and the optimal utilization of land use entitlements. https://cwf.ca/wp-content/uploads/2015/11/CWF_GoingForGold5_EfficientCities_MAR2009.pdf. Accessed 18 May 2020

Bowler DE, Buyung-Ali L, Knight TM, Pullin AS (2010) Urban greening to cool towns and cities: a systematic review of the empirical evidence. Landsc Urban Plan 97(3):147–155

Bramley G, Dempsey N, Power S, Brown C, Watkins D (2009) Social sustainability and urban form: evidence from five British cities. Environ Plan A 41(9):2125–2142

Brand S (1994) How buildings learn: what happens after they're built. Viking, New York

Braubach M, Egorov A, Mudu P, Wolf T, Thompson CW, Martuzzi M (2017) Effects of urban green space on environmental health, equity and resilience. In: Kabisch N, Korn H, Stadler J, Bonn A (eds) Nature-based solutions to climate change adaptation in urban areas. Springer, Cham, pp 187–205

Brinklow A (2010) Transit-oriented development: a policy implementation strategy. Master of urban planning thesis. McGill University, Montreal

Brownett T, Evans O (2019) Finding common ground: the conception of community arts festivals as spaces for placemaking. Health Place 61:102254

Brundtland GH, Khalid M, Agnelli S, Al-Athel S, Chidzero B (1987) Our common future. UN, New York, p 8

Building and Construction Authority (2019) Regulatory requirements for new buildings and existing buildings undergoing major additions and alterations (A&A). https://www1.bca.gov.sg/regulatory-info/legislation-on-environmental-sustainability-for-buildings/regulatory-requirements-for-new-buildings-and-existing-buildings-undergoing-major-additions-and-alterations-(a-a). Accessed 04 Feb 2020

Burden D (2001) Building communities with transportation. https://www.lgc.org/wordpress/docs/freepub/community_design/presentations/dan_burden_2001_pres.pdf. Accessed 17 May 2020

Bureau of Transportation Statistics (2017) National household travel survey daily travel quick facts. https://www.bts.gov/statistical-products/surveys/national-household-travel-survey-daily-travel-quick-facts. Accessed 19 May 2020

Caliskan O, Mashhoodi B (2017) Urban coherence: a morphological definition. Urban Morphol 21(2):123–141

Calthorpe P (1993) The next American Metropolis. Princeton Architectural Press, New York
Carlisle N, Van Geet O, Pless S (2009) Definition of a 'zero net energy' community. https://www.osti.gov/biblio/969716%20Carlisle%20et%20al.%202009. Accessed 06 Apr 2020
Carswell A (2012) The encyclopedia of housing, 2nd edn. SAGE, Los Angeles
Cervero R (2009) Transport infrastructure and global competitiveness: balancing mobility and livability. Ann Am Acad 626:210–224
Cervero R, Guerra E, Al S (2017) Urban recalibration. In: Beyond mobility. Island Press, Washington, DC, pp 1–14
Chamie J (2020) World population: 2020 overview. YaleGlobal online. https://yaleglobal.yale.edu/content/world-population-2020-overview. Accessed 12 Apr 2020
Chandra S (n.d.) Flexible pavement versus rigid pavement. https://www.nbmcw.com/tech-articles/roads-and-pavements/36977-flexible-pavement-versus-rigid-pavement.html. Accessed 11 Feb 2020
Chapman D (2018) Urban design of winter cities: winter season connectivity for soft mobility. Dissertation, Lulea University of Technology
Chen S (2009) Integrating hedgerow into town planning: a framework for sustainable residential development. Int J Human Soc Sci 5(5):293–302
Cheyne C, Imran M (2015) Shared transport: reducing energy demand and enhancing transport options for residents of small towns. Energy Res Soc Sci 18:139–150
Chow WTL, Brazel AJ (2012) Assessing xeriscaping as a sustainable Heat Island mitigation approach for a Desert City. Build Environ 47:170–181
Cintron PA (2014) Chopin mazurkas and its influence on polish nationalism. Dissertation, Liberty University
Cohen S (2018) Defining the sustainable city. Columbia University Press, New York
Collarte N (2012) The Woonerf concept: rethinking a residential street in Sommerville. http://www.solaripedia.com/files/1325.pdf. Accessed 8 Mar 2020
Collins R, Schaafsma M, Hudson MD (2017) The value of green walls to urban biodiversity. Land Use Policy 64:114–123
Commission of the European Committees (2007) Green paper: towards a new culture for urban mobility. European Union, Brussels
Connett P (2013) The zero waste solution: untrashing the planet one community at a time. Chelsea Green Publishing, White River Junction
Coupland A (2013) Reclaiming the city: mixed use development. Taylor and Francis, Hoboken
Davies WKD (2015) Winter cities. In: Davies WKD (ed) Theme cities: solutions for urban problems. Springer, Dordrecht, pp 277–310
Dawson AH (2018) An update on data governance for Sidewalk Toronto. Sidewalk Labs. https://www.sidewalklabs.com/blog/an-update-on-data-governance-for-sidewalk-toronto/#:~:text=Sidewalk%20Labs%20believes%20an%20independent,urban%20data%20originating%20in%20Quayside. Accessed 25 Apr 2020
de la Salle J, Holland M (2010) Agricultural urbanism: handbook for building sustainable food systems in 21st century cities. Green Frigate Books, Sheffield
Ville de Luxembourg (2019) Introduction of free public transport in Luxembourg. https://www.vdl.lu/en/getting-around/bus/tickets/free-travel. Accessed 04 Feb 2020
DeKay M (2010) Daylighting and urban form: an urban fabric of light. J Architect Plann Res 27(1):35–56
Deng T, Nelson JD (2011) Recent developments in bus rapid transit: a review of the literature. Transplant Rev 31(1):69–96
Dittmar H, Ohland G (eds) (2004) The new transit town: best practices. The New Transit Town. Island Press, Washington, DC
Dumbaugh E, Li W (2011) Designing for the safety of pedestrians, cyclists, and motorists in urban environments. J Am Plann Assoc 77(1):69–88

Dutzik T, Madsen T, Baxandall P (2013) A new way to go: the transportation apps and vehicle-sharing tools that are giving Americans the freedom to drive less. https://uspirg.org/sites/pirg/files/reports/A%20New%20Way%20to%20Go%20vUS1_1.pdf. Accessed 22 Apr 2020
Elmaghraby A, Losavio M (2014) Cyber security challenges in smart cities: safety, security, and privacy. J Adv Res 5(4):491–497
European Heart Network (2008) European Heart Network response to the European Commission on the "Green Paper Towards a new culture for urban mobility." http://www.ehnheart.org/publications-and-papers/responses-to-consultations.html?start=20. Accessed 18 May 2020
European's Cyclist Federation (2009) Future cities are cycling cities! https://ecf.com/sites/ecf.com/files/Future-cities-are-cycling-cities.pdf. Accessed 18 May 2020
Ewing R (2013) Eight qualities of pedestrian- and transit-oriented design. Urban Land Institute and American Planning Association, Washington, DC
Ewing R, Handy S (2009) Measuring the unmeasurable: urban design qualities related to walkability. J Urban Des 14(1):65–84
Farris J (2001) The barriers to using infill development to achieve smart growth. Hous Policy Debate 1:1–30
Feehan D, Feit M (2006) Making business districts work. Routledge, New York
Forest City (2019) Masterplan. https://forestcitycgpv.com/why-forest-city/masterplan/. Accessed 04 Feb 2020
Freixas C, Abbott M (2018) Segregation by design. Springer Nature, Cham
Freymann G (2014) About the CHTA. https://www.chta.ca/about-us.html. Accessed 18 May 2020
Friedman A (2002) Planning the new suburbia: flexibility by design. UBC Press, Vancouver
Friedman A (2005) Homes within reach. Wiley, Hoboken, NJ
Friedman A (2007) Sustainable residential development: planning and design for Green neighbourhoods. McGraw-Hill, New York
Friedman A (2010) A place in mind: the search for authenticity. Véhicule Press, Montréal
Friedman A (2012a) Fundamentals of sustainable dwellings. Island Press, Washington, DC
Friedman A (2012b) Open spaces. In: Town and terraced housing: for affordability and sustainability. Routledge, Abingdon, pp 156–179
Friedman A (2013) Innovative houses: concepts for sustainable living. Laurence King, London
Friedman A (2014) Planning small and mid-sized towns: designing and retrofitting for sustainability. Routledge, New York
Friedman A (2015a) Fundamentals of sustainable neighbourhoods. Springer, Cham
Friedman A (2015b) Planning sustainable neighbourhoods. Springer, Cham
Friedman A (2017) Designing sustainable communities. Bloomsbury, London
Friedman A (2018a) Fundamentals of sustainable urban renewal in small and mid-sized towns. Springer, Cham
Friedman A (2018b) Neighbourhood: designing a livable community. Véhicule Press, Montreal
Garrard GE, Williams NS, Mata L, Thomas J, Bekessy SA (2018) Biodiversity sensitive urban design. Conserv Lett 11(2):e12411
Gatien A, Mas Baghaie A (2019) Improving active transportation and public transit integration: a guidebook for policy and planning. https://www.tcat.ca/wp-content/uploads/2019/06/Active-Transportation-and-Public-Transit-Integration-web-3.pdf. Accessed 23 Mar 2020
Gehl J (2010) Cities for people. Island Press, Washington, DC
Gehl J (2011) Life between buildings. Island Press, Washington, DC
Gehl J (2017) In the last 50 years architects have forgotten what a good human scale is. https://www.archdaily.com/877602/jan-gehl-in-the-last-50-years-architects-have-forgotten-what-a-good-human-scale-is. Accessed 22 Jan 2020
Gehl Architects (2014) World class streets: remaking New York City's public realm. Gehl Architects, New York
Gehl Architects (2020) Masterplanning frameworks. Gehl Architects, New York
Geller A (2011) Smart growth: a prescription for liveable cities. Am J Public Health 93:1410–1415

Government of Canada (2014) Active transportation. https://www.canada.ca/en/public-health/services/being-active/active-transportation.html. Accessed 22 Apr 2020
Grant J (2002) Mixed use in theory and practice: Canadian experience with implementing a planning principle. J Am Plann Assoc 68(1):71–84
Hagen K, Stiles R (2010) Contribution of landscape design to changing urban climate conditions. In: Müller N, Werner P, Kelcey JG (eds) Urban biodiversity and design. Blackwell, London, pp 572–592
Hales CM, Carroll MD, Fryar CD, Ogden CL (2020) Prevalence of obesity and severe obesity among adults: United States, 2017–2018. NCHS Data Brief. Available via CDC. https://www.cdc.gov/nchs/products/index.htm. Accessed 18 May 2020
Hall E (2007) Divide and sprawl, decline and fall: a comparative critique of Euclidean zoning. Univ Pittsburgh Law Rev 68(4):915–952
Hall P (2008) Urban renaissance, urban villages, smart growth: find the differences. In: Haas T (ed) New urbanism and beyond: designing cities for the future. Rizzoli, New York
Han B, Zhang L, Ou J (2017) Photocatalytic concrete. In: Han B, Zhang L, Ou J (eds) Smart and multifunctional concrete toward sustainable infrastructures. Springer, Singapore, pp 299–311
Hand C, Riley F, Harris P, Singh J, Rettie R (2009) Online grocery shopping: the influence of situational factors. Eur J Mark 43(9/10):1205–1219
Hashem I, Chang V, Anuar N, Adewole K, Yaqoob I, Gani A, Ahmed E, Chiroma H (2016) The role of big data in the smart city. Int J Inf Manag 36(5):748–758
Hein HS (2006) Public art: thinking museums differently. Rowman Altamira Press, Lanham
HERE Mobility (2020) Shared transport: moving toward a common goal. https://mobility.here.com/shared-transport-moving-towards-common-goal. Accessed 08 Apr 2020
Hersperger A (2006) Spatial adjacencies and interactions: neighbourhood mosaics for landscape ecological planning. Landsc Urban Plan 77(3):227–239
Hodges T (2010) Public transportations role in responding to climate change. Diane Publishing, Darby
Holden R, Liversedge J (2011) Construction for landscape architecture. Laurence King, London
Howard E (1902) Garden cities of to-morrow. Sonnenschein., London
Howell SC (1972) Privacy as an expression of human territoriality. In: Palastan L, Carson D (eds) Spatial behavior of older people. University of Michigan Press, Ann Arbor, MI
Institute for Transport and Development Policy (ITDP) (n.d.) What is BRT? https://www.itdp.org/library/standards-and-guides/the-bus-rapid-transit-standard/what-is-brt/. Accessed 22 Apr 2020
Instituto Distrital de Recreación y Deporte (2019) Ciclovía Bogotana. https://www.idrd.gov.co/ciclovia-bogotana. Accessed 04 Feb 2020
Islam S (2011) Traditional urban planning approaches and Sustainable City. Open House Int 36(2):15–23
Ismail WHW, Ching LH (2016) Back lanes as social spaces in Chinatown, Kuala Lumpur. Env Behav Proc J 1(3):293–299
Jackson L (2003) The relationship of urban design to human health and condition. Landsc Urban Plan 64(4):191–200
Jacobs J (1961) The death and life of great American cities. Vintage Books, New York
Jacobs J (1969) The economy of cities. Random House, New York
Jacobsen PL, Racioppi F, Rutter H (2009) Who owns the roads? How motorised traffic discourages walking and bicycling. Inj Prev 15(6):369–373
Jaffe E (2016) How to keep cyclists riding even in the frigid snowy winter. https://www.citylab.com/transportation/2016/01/winter-bike-riding-seasonal-cycling/426960/. Accessed 05 May 2020
Johnson M (2001) Environmental impacts of urban sprawl: a survey of the literature and proposed research agenda. Environ Plan A 33(4):717–735
Kamarulzaman N, Hashim SZ, Hashim H, Saleh AA (2014) Green roof concepts as a passive cooling approach in tropical climate-an overview. E3S web of conferences 3:01028

Kaswan V, Choudhary M, Kumar P, Kaswan S, Bajya P (2019) Green production strategies. Elsevier, Amsterdam
Katzev R (2003) Car sharing: a new approach to urban transportation problems. Anal Soc Issues Public Policy 3(1):65–86
Kaw JK, Lee H, Wahba S (2020) The hidden wealth of cities: creating, financing and managing public spaces. The World Bank Group, Washington, DC
Kay J (1997) Asphalt nation. University of California Press, Berkeley
Kim J (2016) Achieving mixed income communities through infill? The effects of infill housing on neighborhood income diversity. J Urban Aff 38(2):280–297
King D (2014) Three big challenges for planning multi-modal cities. https://www.citylab.com/design/2014/10/3-big-challenges-for-planning-multi-modal-cities/381254/. Accessed 23 Mar 2020
Kiper T (2013) Role of ecotourism in sustainable development. In: Özyavuz M (ed) Advances in landscape architecture. Namık Kemal University, Turkey, pp 773–802
Kong F, Yin H, Nakagoshi N, Zong Y (2010) Urban green space network development for biodiversity conservation: identification based on graph theory and gravity modeling. Landsc Urban Plan 95(1–2):16–27
Kostof S (1991) The city shaped: urban patterns and meanings through history. Thames & Hudson, London
Krygsman S, Dijst M, Arentze T (2004) Multimodal public transport: an analysis of travel time elements and the interconnectivity ratio. Transp Policy 11(3):265–275
Kuo FE (2003) Social aspects of urban forestry: the role of arboriculture in a healthy social ecology. J Arboric 29(3):148–155
Lachapelle U, Noland RB (2012) Does the commute mode affect the frequency of walking behavior? The public transit link. Transp Policy 21:26–36
Larsen K, Gilliland J (2009) A farmers' market in a Food Desert: evaluating impacts on the Price and availability of healthy food. Health Place 15(4):1158–1162
Lepczyk CA, Aronson MFJ, Evans KL, Goddard MA, Lerman SB, MacIvor JS (2017) Biodiversity in the city: fundamental questions for understanding the ecology of urban green spaces for biodiversity conservation. BioScience 67(9):799–807
Lindfield M, Steinberg F (2012) Green Cities. Asian Development Bank Institute. ProQuest Ebook Central, https://ebookcentral.proquest.com/lib/mcgill/detail.action?docID=3110769. Accessed 04 Feb 2020
Litman T (2017) Introduction to multi-modal transportation planning: principles and practices. Victoria Transport Policy Institute, Victoria, BC
Logan JR (1976) Industrialization and the stratification of cities in suburban regions. Am J Sociol 82(2):333–348
Loukaitou-Sideris A, Levy-Storms L, Brozen M (2014) Placemaking for an aging population. https://www.lewis.ucla.edu/wp-content/uploads/sites/2/2015/04/Seniors-and-Parks-8-28-Print_reduced.pdf. Accessed 19 May 2020
Low N (2003) The active city. Urban Policy Res 21(1):5–7
Luk N, Olszewski P (2003) Integrated public transport in Singapore and Hong Kong. Road Transp Res 12(4):41–51
Lynch K (1960) The image of the City. MIT Press, Cambridge
MacIvor JS (2014) Building height matters: nesting activity of bees and wasps on vegetated roofs. Israel Journal of Ecology & Evolution 62(1–2):88–96
Mahmoud AHA, El-Sayed MA (2011) Development of sustainable urban green areas in Egyptian new cities: the case of El-Sadat City. Landsc Urban Plan 101(2):157–170
Manso M, Castro-Gomes J (2015) Green Wall systems: a review of their characteristics. Renew Sustain Energy Rev 41:863–871
Martens K (2006) Promoting bike-and-ride: the Dutch experience. Transp Res A 41(4):326–338
Mayrand F, Clergeau P (2018) Green roofs and green walls for biodiversity conservation: a contribution to urban connectivity? Sustainability 10(4):985

McConnell V, Wiley K (2010) Infill development: perspectives and evidence from economics and planning. Resour Future 10:13
Mckenzie S (2004) Social sustainability: towards some definitions. Hawke Research Institute, University of South Australia, Magill, South Australia
Mecredy G, Pickett W, Janssen I (2011) Street connectivity is negatively associated with physical activity in Canadian youth. Int J Environ Res Public Health 8(8):3333–3350
Meltzer L (2014) Consideration of climate change in the design of parks and open spaces. https://www.hybridparks.eu/wp-content/uploads/downloads/2014/08/Climate_Change_NRW_en.pdf?fbclid=IwAR1ttF471AUzOxe87-LbyX3cNB8lPiSZt3zyOz1YCuKylOWiYMxhP5y6BhM. Accessed 26 Apr 2020
Mendler S, Odell W (2000) The HOK guidebook to sustainable design. Wiley, Hoboken, NJ
Mertens S (2002) Wind energy in urban areas: concentrator effects for wind turbines close to buildings. Refocus 3(2):22–24
MetLife (2011) Market survey of long-term care costs. https://www.aarp.org/content/dam/aarp/livable-communities/old-learn/health/the-metlife-market-survey-of-nursing-home-assisted-living-adult-day-services-and-home-care-costs-2011-aarp.pdf. Accessed 11 May 2020
Miller M (1992) Raymond Unwin: garden cities and town planning. Burns & Oates, London
Mohanty S, Choppali U, Kougianos E (2016) Everything you wanted to know about smart cities: the internet of things is the backbone. IEEE Consum Electr Mag 5(3):60–70
Morris S, Fawcett G, Brisebois L, Hughes J (2018) a demographic, employment and income profile of Canadians with disabilities aged 15 years and over, 2017. https://www150.statcan.gc.ca/n1/pub/89-654-x/89-654-x2018002-eng.htm. Accessed 12 May 2020
National Association of City Transportation Officials (NACTO) (2012) Urban street design guide. https://nacto.org/publication/urban-street-design-guide/. Accessed 11 Feb 2020
National Association of City Transportation Officials (NACTO) (2014) Conventional bike lanes. https://nacto.org/publication/urban-bikeway-design-guide/bike-lanes/conventional-bike-lanes/#. Accessed 23 Apr 2020
National Association of City Transportation Officials (NACTO) (2016) Transit street design guide. https://nacto.org/publication/transit-street-design-guide/. Accessed 23 Mar 2020
National Association of City Transportation Officials (NACTO) (2018) Guidelines for the regulation and management of shared active transportation. https://nacto.org/wp-content/uploads/2018/07/NACTO-Shared-Active-Transportation-Guidelines.pdf. Accessed 08 Apr 2020
Nelson A (2011) Coloring inside the lanes: art that creates community. https://grist.org/cities/2011-12-02-coloring-inside-the-lanes-art-community/. Accessed 15 May 2020
North A (2013) Operative landscapes: building communities through public space. Birkäuser, Basel
Numbers MJ (1995) Siting a house. Fine Homebuild 93:40–45
OECD (2013) Green growth in cities. OECD Green Growth Studies. OECD, Paris. Accessed 04 Feb 2020. https://doi.org/10.1787/9789264195325-en
Oldenburg R (1999) The great good place: cafes, coffee shops, bookstores, bars, hair salons, and other hangouts at the heart of a community. Da Capo Press, Cambridge
Orbasli A (2000) Tourists in historic towns: urban conservation and heritage management. E&FN Spon, New York and London
Orkustofnun (n.d.) Snow melting. https://nea.is/geothermal/direct-utilization/snow-melting/. Accessed 17 May 2020
Oyez (n.d.) Village of Euclid v. Ambler Realty Company. https://www.oyez.org/cases/1900-1940/272us365. Accessed 1 Feb 2020
Pacheco R, Ordóñez J, Martínez G (2012) Energy efficient design of building: a review. Renew Sustain Energy Rev 16(6):3559–3573
Pantoja AH (1983) Site planning for low-rise housing, with special reference to northern climates. Unpublished M. arch report, McGill University, School of Architecture

Parolek D, Parolek K, Crawford P (2008) Form based codes: a guide for planners, urban designers, municipalities, and developers. Wiley, Hoboken, NJ
Parris KM, Amati M, Bekessy SA, Dagenais D, Fryd O, Hahs AK, Hes D, Imberger SJ, Livesley SJ, Marshall AJ, Rhodes JR (2018) The seven lamps of planning for biodiversity in the city. Cities 83:44–53
Pauleit S, Zölch T, Hansen R, Randrup TB, van den Bosch CK (2017) Nature-based solutions and climate change–four shades of green. In: Kabisch N, Korn H, Stadler J, Bonn A (eds) Nature-based solutions to climate change adaptation in urban areas. Springer, Cham, pp 29–49
Perkl RM (2016) Geodesigning landscape linkages: coupling GIS with wildlife corridor design in conservation planning. Landsc Urban Plan 156:44–58
Pfeffer J (2010) Building sustainable organizations: the human factor. Acad Manag Perspect 24:34. https://doi.org/10.5465/amp.24.1.34
Pradhan B (2017) Spatial modelling and assessment of urban form. Springer, Cham
Pressman NEP (1996) Sustainable winter cities: future directions for planning, policy and design. Atmos Env 30(3):521–529
Pressman, N (2004) Shaping cities for winter: climatic comfort and sustainable design. Prince George, Winter Cities Association
Priyanka BA, Sarang G, Ravi Shankar AU (2019) Evaluation of Superpave mixtures for perpetual asphalt pavements. Road Mater Pave Des 20(8):1952–1965
Pucher J, Buehler R (2008) Making cycling irresistible: lessons from the Netherlands, Denmark and Germany. Transplant Rev 28(4):495–528
Putnam RD (2000) Bowling alone: the collapse and revival of American community. Simon & Schuster, New York
Radwan AH, Morsy AAG (2017) The importance of integrating street furniture in the visual image of the city. Int J Mod Eng Res 9(2)
Rao A (2018) Our cities are designed for loneliness. https://www.vice.com/en_ca/article/kzvzpv/our-cities-are-designed-for-loneliness-v25n4. Accessed 18 May 2020
Rathnayake U, Anwar AF (2019) Dynamic control of urban sewer systems to reduce combined sewer overflows and their adverse impacts. J Hydrol 579:124150
Reith T (2019) Embracing and Edmonton winter takes layers of clothing—and a leap of faith. https://www.cbc.ca/news/canada/edmonton/edmonton-winter-embrace-1.5401015. Accessed 05 May 2020
Renne JL, Wells JS (2004) Emerging European-style planning in the USA: transit-oriented development. World Transp Policy Pract 10(2):12–24
Rudlin D, Hemani S (2019) Climax city: master planning and the complexity of urban growth. RIBA, London
Ruff J (2007) For sale: the American dream. World History Group, Leesburg, VA
Salingaros N (2000) Complexity and urban coherence. J Urban Des 5(3):291–316
Samanvi Technologies (2020) Unipaver. http://www.unipaver.in/. Accessed 11 Feb 2020
Schoenauer N (2000) 6,000 years of housing. WW Norton, New York
Schraad-Tischler D, Seelkopf L (2015) Concept and methodology—sustainable governance indicators. https://www.sgi-network.org/2019/. Accessed 06 Apr 2020
Selbig WR, Buer N (2018) Hydraulic, water-quality, and temperature performance of three types of permeable pavement under high sediment loading Conditions. U.S. Geological Survey. https://doi.org/10.3133/sir20185037
Sharma S (n.d.) Interlocking concrete paver blocks. https://www.nbmcw.com/tech-articles/concrete/4993-interlocking-concrete-paver-blocks.html. Accessed 11 Feb 2020
Shepperson M (2009) Planning for the sun: urban forms as a Mesopotamian response to the sun. World Archaeol 41(3):363
Sim D (2019) Soft City: building density for everyday life. Island Press, Washington, DC
Simões Aelbrecht P (2016) 'Fourth places': the contemporary public settings for informal social interaction among strangers. J Urban Des 21(1):124–152

Sinha S (2012) Living longer, living well: recommendations to inform a seniors strategy in Ontario. http://www.health.gov.on.ca/en/common/ministry/publications/reports/seniors_strategy/docs/seniors_strategy_report.pdf. Accessed 14 May 2020
Song Y, Knaap G (2004) Measuring urban form: is Portland winning the war on sprawl? J Am Plann Assoc 70(2):210–225
Southworth M (1997) Walkable suburbs? An evaluation of neotraditional communities at the urban edge. J Am Plann Assoc 63(1):28–44
Squires G (ed) (2002) Urban sprawl: causes, consequences, and policy responses. The Urban Institute Press, Washington, DC
Statistics Canada (2018) The shift to smaller households over the past century. https://www150.statcan.gc.ca/n1/pub/11-630-x/11-630-x2015008-eng.htm. Accessed 01 Apr 2020
Statistics Canada (2019) Analysis: population by age and sex. https://www150.statcan.gc.ca/n1/pub/91-215-x/2018002/sec2-eng.htm. Accessed 01 Apr 2020
Steinberg L (2015) Woonerf: inclusive and livable Dutch street. https://www.humankind.city/post/woonerf-inclusive-and-livable-dutch-street. Accessed 08 Mar 2020
Stout M, Collins D, Stadler SL, Soans R, Sanborn E, Summers RJ (2018) "Celebrated, not just endured:" rethinking winter cities. Geogr Compass 12(8):e12379
Swingland IR (2013) Biodiversity, definition of. In: Levin SA (ed) Encyclopedia of biodiversity, 2nd edn. Elsevier, Amsterdam, pp 399–410
Tam T (2017) The chief public Health Officer's report on the state of public health in Canada 2017: designing healthy living. Public Health Agency of Canada. https://www.canada.ca/en/public-health/services/publications/chief-public-health-officer-reports-state-public-health-canada/2017-designing-healthy-living.html?wbdisable=true. Accessed 03 May 2020
Thompson H (2018) Combining E-bikes and bikeshares is urban alchemy: a winning transit solution for cities. https://www.forbes.com/sites/energyinnovation/2018/01/08/combining-e-bikes-and-bikeshares-is-urban-alchemy-a-winning-transit-solution-for-cities/#1ca9dbed3e09. Accessed 08 Apr 2020
Thompson H, Kennan H (2017) As transportation costs, emissions grow, electric bikes offer an efficient alternative. https://www.forbes.com/sites/energyinnovation/2017/07/06/as-transportation-costs-emissions-grow-electric-bikes-offer-an-efficient-alternative/#5a1a0e2a305a. Accessed 08 Apr 2020
Tregoning H, Agyeman J, Shenot C (2002) Sprawl, smart growth, and sustainability. Local Env 7(4):341–347
United Nations (2006) Population aging 2006. https://www.un.org/en/development/desa/population/publications/pdf/ageing/wallchart-ageing2006.pdf. Accessed 11 May 2020
United Nations (2011) World population prospects: the 2010 revision, volume I: comprehensive tables. https://www.un.org/en/development/desa/population/publications/pdf/trends/WPP2010/WPP2010_Volume-I_Comprehensive-Tables.pdf. Accessed 18 May 2020
United Nations (2014) World's Population Increasingly Urban with More than Half Living in Urban Areas. https://www.un.org/development/desa/en/news/population/world-urbanization-prospects.html. Accessed 05 May 2020
United Nations (2017) The New Urban Agenda. http://habitat3.org/wp-content/uploads/NUA-English.pdf. Accessed 04 Feb 2020
United Nations (2018) Department of economic and social affairs. https://www.un.org/development/desa/en/news/population/2018-revision-of-world-urbanization-prospects.html. Accessed 03 Mar 2020
United Nations (2019) World urbanization prospects: the 2018 revision. United Nations, New York
Uslu A, Shakouri N (2013) Urban landscape design and biodiversity. In: Özyavuz M (ed) Advances in landscape architecture. Namık Kemal University, Turkey
Van der Biest K, Meire P, Schellekens T, D'hondt B, Bonte D, Vanagt T, Ysebaert T (2020) Aligning biodiversity conservation and ecosystem services in spatial planning: focus on ecosystem processes. Sci Total Environ 712:136350

Van Mierlo J, Maggetto G (2007) Fuel cell or battery: electric cars are the future. Fuel Cells 7(2):165–173

Watsky R (2018) The problems with Euclidean zoning. http://sites.bu.edu/dome/2018/07/19/the-problems-with-euclidean-zoning/. Accessed 01 Feb 2020

Wells W (ed) (2007) Blueprint for greening affordable housing. Island Press, Washington, DC

Wilkerson J (2018) Biomass over coal: burning different carbon to mitigate climate change. http://sitn.hms.harvard.edu/flash/2018/biomass-over-coal-burning-different-carbon-to-mitigate-climate-change/. Accessed 18 May 2020

Wintercity Strategy (2018) For the love of winter: strategy for transforming Edmonton into a world-leading winter city. https://www.edmonton.ca/city_government/documents/PDF/COE-WinterCity-Love-Winter-Summary-Report.pdf. Accessed 05 May 2020

Winters M, Brauer M, Setton EM, Teschke K (2010) Built environment influences on healthy transportation choices: bicycling versus driving. J Urban Health 87(6):969–993

World Health Organization (WHO) (2009) WHO global network of age-friendly cities. https://www.who.int/ageing/Brochure-EnglishAFC9.pdf. Accessed 13 May 2020

World Health Organization (WHO) (2015) World report on aging and health. WHO Library Cataloguing. https://apps.who.int/iris/bitstream/handle/10665/186463/9789240694811_eng.pdf;jsessionid=D2CD5F07B0FA2DE0861F9681C920B12D?sequence=1. Accessed 13 May 2020

Yücel GF (2013) Street furniture and amenities: designing the user-oriented urban landscape. In: Özyavuz M (ed) Advances in landscape architecture. Namık Kemal University, Turkey

Zanella A, Bui N, Castellani A, Vangelista L, Zorzi M (2014) Internet of things for smart cities. IEEE Internet Things J 1(1):22–32

Zhang Z, Meerow S, Newell JP, Lindquist M (2019) Enhancing landscape connectivity through multifunctional green infrastructure corridor modeling and design. Urban For Urban Green 38:305–317

Zhivov A, Case M, Liesen R, Kimman J, Broers W (2014) Energy master planning towards net-zero communities and campuses. ASHRAE Trans 120(1):114–129

Zhou B, Li W, Chan K, Cao Y, Kuang Y, Liu X, Wang X (2016) Smart home energy management systems: concept, configurations, and scheduling strategies. Renew Sustain Energy Rev 61:30–40

Zimmer C, Fox B, Dhalla S, Walters M (2012) Low impact development discussion paper. ICF Markbek. https://sustainabletechnologies.ca/app/uploads/2014/09/LID-Discussion-Paper_Nov-2012.pdf. Accessed 18 May 2020

zu Ermgassen SOSE, Utamiputri P, Bennun L, Edwards S, Bull JW (2019) The role of "no net loss" policies in conserving biodiversity threatened by the global infrastructure boom. One Earth 1(3):305–315

Index

A
Accessibility, 197, 199
Active mobility, 150–153, 171, 173, 174, 177, 179
Active modes of transportation, 77
Active transportation, 98, 145, 147, 151, 152, 188
Adaptable building techniques, 113
  add-in method, 113
  flexible layout, 110
  global population ages, 113
  horizontal expansion, 113
  residual spaces, 111
Adaptable communities, 108
Affordability gap, 9, 197
Affordable housing, 87, 98, 101, 155, 161
Age-friendly communities, 209
Aging population, 6
Alternative standards
  active and public transit, 163
  amenities, 164
  designing streets, 163
  ground cover, 165, 167, 168
  human scale, 163
  lanes, 165
  narrower streets, 164, 165
  natural environment, 164
  paths, 165
Amenities, 67, 68, 70, 147
American Public Transportation Association (APTA), 198
American Society of Heating, 125
Architectural idioms, 115, 116
Arid climates, 60
Artistic expression, 297, 301
Asian Development Bank, 10
Automobile-dependent growth, 95

B
Back lanes, 288, 291, 292
Balancing higher densities, 54
Bay Area Rapid Transit (BART), 159
Benny Farm community, 92
Benny Farm project, 106
Benny Farms, 92
Bike sharing, 188, 197, 199
Biodiversity
  biological interactions, 247
  built environment, 245
  cycles, 247
  definition, 245
  genetic diversity, 247
  green roofs, 250
  green walls, 250–252
  metapopulations, 247
  natural disturbance regimes, 245
  pollinators, 247
  principles of planning, 247, 249
  retooling cities, 249
  urban landscape, 247
Biological interactions, 247
Biomass energy, 131
Biotic homogenization, 245
Brownfield sites, 98
Building heights, 119

A. Friedman, *Fundamentals of Sustainable Urban Design*,
https://doi.org/10.1007/978-3-030-60865-1

Buildings' forms
  active mobility, 71
  commercial and residential activity, 69
  function, 67
  mixed land uses, 67
  mixed-use neighborhoods, 69
  mobility, 70
  semi-enclosed space, 71
Buildings' forms and open spaces
  community members, 73
  mixed-use areas, 73
  sustainable development, 73
  urban development, 71
Built-infrastructure, 243
Bus rapid transit (BRT), 151
Business practices, 18

C
Carbon dioxide emissions, 7
Carbon footprint, 26
Car-centric design, 75
Car-free environments, 181, 183
Car park controls, 190
Carpooling, 198
Car sharing programs, 138
Cities, 291, 292
City planning, 53
City squares, 73
Climate change, 6, 10, 257, 259
Climate-responsive approaches, 216
Climate sensitive design, 217, 218
Cohesive space, 301
Communal spaces, 71, 73
Communal waste collection, 133
Community, 145
Community gardens, 280, 281
Compact cities, 48
Complete place, 70
Congestion, 190, 191
Connected green networks, 230
Connectivity, 107, 158
Consumer electronics, 140
Contemporary urban challenges
  aging population, 6
  economic activities, 8
  environmental domain, 7
  environmental issues, 8
  social transformation, 6
Contour patterns, 57, 58
Contra Costa Centre Village, 159
Corridors, 225, 228
Cost-effective transportation, 157
Covered sidewalks, 215
Cradle-to-cradle cycle assessment, 19
Created cities, 38
Crosswalks, 174
Crystal palace, 40
Cultural preservation, 118
Cultural sustainability, 17, 18
Culture, 295
Cycling, 188, 189, 213
Cycling culture, 78
Cyclists, 171, 172, 177

D
Demographic changes, 6
Density, 158, 160
Density indices, 50
Digital city
  feasibility, 137
  human and financial capital, 137
  IoT, 136
  smart phones, 135
  urban infrastructure, 135
Digital technologies, 139
Digital tools
  online grocery shopping, 138
  public transit users, 138
  support services, 138
  transport data, 138
District heating system, 129
Diversity, 158
Diversity of land uses, 86
Divided bike paths, 172
Dockless system, 201
Dockside Green project, 132
Dutch woonerf experience, 183, 185
Dwellings, 101

E
Ecological footprint, 37, 235
Economic and social sustainability, 56
Economic incentives, 153
Economic stress, 187
Economic sustainability, 18, 206, 209
Economy of Cities, 37
Ecosystem cycles, 247
Ecosystem qualities, 245
Ecosystems, 236, 241, 242
Edible landscapes, 279, 281
Education, 207
Electric bikes, 198–200
Endangering species, 245

Energy distribution, 130
Energy-efficient, 156
Energy generation, urban areas
  conservation movement, 125
  fossil fuels, 127
  net-zero community, 125
  solar gain, 129
  solar panels, 127, 128
  train system, 128
Energy production, 127
Environmental site assessment (ESA), 98
Environmental sustainability, 16
Euclidean zoning, 43, 85, 93
Exclusionary regulations, 217
Existing natural features, 230
  ecological footprint, 235
  human activity, 235
  hyper-urbanization, 235
  integration, 236
  quality of life, 235
  typology, 237, 238
  urban design methods, 238–243
Expandable dwellings, 111, 112

F
Farmers' markets, 282, 284
Fauna, 236, 237, 242
Features, 210
Fertilizers, 257
Fit Urban Forms, 53, 54
Flexible home construction, 113
Flexible planning
  and building strategies, 105
Flexible urban outgrowth, 108
Floor Area Ratio (FAR), 50
Flora and fauna, 57, 236, 237, 240–242
Food exchanges, 282, 284
Food system, 279
Food waste, 279
Form-based codes, 120, 122
Free Wi-Fi service, 136
Freeway service roads, 177

G
Garden Cities of Tomorrow, 40
Gehl Master planning Frameworks, 34
Genetic variation, 245
Geo-fencing technologies, 201
Gini–Simpson index, 116, 118
Global food system, 277
Graffiti and street art, 292
Green belt, 226, 227
Green city thinking
  building cities, 10
  environmental commitments and movements, 10
  investment, 11, 12
Green growth, 10
Green modes of transit, 150–153
Green networks, 240
Green roofs, 249, 250
Green spaces, 115, 116, 268
Green technologies, 150
Green walls, 250–252
Greenfield development, 110
Greenhouse gas emissions, 130, 187
Greenspaces, 279
Greenwashing, 13
Grid layout, 39
Gross density, 49
Grouped parking, 182, 184
Grouping buildings, 71

H
Hammarby's waste management system, 131
Harmony and diversity, 116
Herbicides, 257
Heritage sites, 118
Hierarchical characterization, 110
Hierarchical importance, 108
Hierarchical system, 172
Hierarchy configuration, 226, 228
High density, 51
Higher densities, 145, 148, 149
High-pressure sodium (HPS), 254
Homogeneity
  coherence, 118
  evolutionary design, 120
  function and importance, 119
  infill study, 119
  modern experience, 118
Hot Mix Asphalt (HMA), 166
Housing, 86, 155–161
Howard, Ebenezer, 40
Human scale, 79, 81
  design, 81
  mind, 79
  neighborhood, 79
  post-war planning practices, 79
  urban designers, 79

I
Imageability, 77
Improved health, 207
Inclusive, 205, 206, 209, 211, 212
Independence and integration, 206
Infill building process, 107
Infill housing projects, 96, 100
Infill projects, 95, 98, 101, 113
  barriers, 98
  benefit, 98
  form, 98
  legal and political challenges, 99
Installing artwork, 297
Integrated curved, 184
Interactive nature, 298
Interconnected digital infrastructure, 135
Interlocking Concrete Block Pavement (ICBP), 167
Internet of Things (IoT), 137
Involvement, 298, 306

K
Kostof's typology, 40

L
Land and urban form, 45
Landscaping, 231, 232
Land use, 157
Land use allocation, 33
Land-use distribution, 40
Land-use management, 145
Land-use planning, 108
Land-use policies, 148
Layering of building, 69
Letchworth, 40
Life cycle approach, 24
Lifestyle, 268, 271, 275
Light rail system, 145
Light rail transit, 190
Lights and light-emitting diodes (LEDs), 254
Linear-development plan, 145
Links, 225
Livable places, 171
Live-work housing, 98
Living wall systems (LWS), 250
Local opposition, 98
Local wind patterns, 129
Locating amenities, 147, 149
Lot size, 119
Low-density projects, 51
Low impact development (LID), 150, 249

M
Macro land-use planning, 148
Macro-level coherence, 115
Main streets, 73
Master plan, 29
  approaches and strategies, 33
  approaching, 29
  development context, 31
  land use, 33
  public investment, 31
  regeneration, 31
  urban contexts, 30, 31
  urban growth, 32
Mental health, 279
Micro-climates, 215, 218
Mixed communities, 86
Mixed housing types, 159
Mixed land use, 90, 158
Mixed-use, 122
Mixed-use buildings, 78
Mixed-use developments, 268, 271
Mixed-use panning, 86, 87
  businesses benefit, 89
  designers, 85
  dwelling types, 85
  economic investment, 89
  green spaces, 87
  segregation, 86
  urban areas, 86
  zoning, 85
Mixed-use urban space, 91
Mixing land uses, 86, 89
Mobility, 145, 148, 149
Mobility and connectivity
  land use, 34
  redevelopment, 36
Motor output, 207
Mulch, 263–265
Multimodal methods, 157
Multimodal transit networks, 187–190
Multimodal transportation, 198

N
Native species, 240
Native wildlife, 238
Natural bumps and bends, 57
Natural drainage system, 60
Natural settings, 150
Neighborhood configuration, 89
Neighborhood density, 50
Net density, 49
Net-zero community, 125

New Urban Agenda, 10
New Urbanism, 45
Night-time light pollution, 253, 254
NIMBYism, 98
Nodes, 225
Nordic cities, 215–217

O
Obesity, 267, 275
Open spaces
  exercise, 274, 275
  green infrastructure, 225
  high-density urban spaces, 226
  integration, 225
  outdoors areas, 270, 271
  play areas, 273
  public health challenges, 267
  recreation, 274, 275
  strategies, 230–232
  typology, 226, 228, 230
  urban development, 225
  urban resilience, 226
  walkability, 268, 270
Optimal infill site, 100
Outdoor living rooms, 287
Outdoors areas, 71
Outdoor spaces, 213, 216, 218–220

P
Parking, 119
Parking spaces, 182, 186
Passive cooling
  environment, 65
  vegetation, 64
  ventilation, 64
  windbreaks, 65
  wind flow, 65
Passive solar gain
  dwelling, 62
  environment, 64
  mesopotamian cities, 61
  natural vegetation, 62
  renewable forms, 62
Pathways, 174
Paved surfaces, 240
Pedestrian access, 107
Pedestrian network, 51
Pedestrian pockets, 90, 91
Pedestrians, 171, 172, 174–177
Performance spaces, 289, 290
Pesticides, 257
Physical abilities, 270
Physical activity, 213, 214
Physical barriers, 214
Placemaking, 264, 266, 287, 289, 292
Planned cities, 38
Planned Unit Development (PUD), 149
Planning methods, 89
Plant species, 279
Play and creativity, 296
Play areas, 271, 273, 274
Playground, 232, 233
Pocket parks, 56, 271, 274
Principle density indices, 49
Psychological attributes, 237
Public art, 287, 295, 297, 298, 302, 305
Public health, 6, 138, 150
Public infrastructure, 149
Public open areas, 268
Public open spaces, 56
Public outdoor spaces, 54
Public places, 115
Public space, 71, 73
Public transit, 181, 183, 190, 268, 270
Public transit networks, 51, 148, 182
Public transit stops, 192–194
Public transportation, 197, 199, 200, 203

R
Radburn, 41, 42
Rain gardens, 279
Rainwater runoff, 231
Range of species, 245
Reduced mobility, 206, 208
  adults, 208
  aging population, 205–207, 209
  challenges, 207
  characteristics, 205
  community planning, 209
  designing cities, 205
  economic sustainability, 206, 209
  integration, 209
  interdependence, 209
  mental health/mood disorders, 207
  mobility, 208
  planning, 206
  safety, 206
  United Nations, 205
  universal design, 209
  urban design strategies, 209–212
  urban environments, 205
Renewable energy sources, 150
Renewable resources, 210
Residential density, 49, 173
Resilient communities, 107
Retail hubs, 89
Road networks, 187

Roadside utilization, 300
Runoff, 258, 260
Runoff drainage, 240
Rural-urban migration, 105
Russell Sage Foundation, 43

S
Safe spaces, 190
Safe streets, 206, 209
Seasonal variability, 239
Self-sustaining system, 22
Sense of community, 70, 73
Sense of place, 183, 185
  benefit city residents, 78
  comfort and community, 77
  compelling case, 77
  identity, 75
  imageability, 77
  mixed-use buildings, 77
  paradigm shift, 75
  social interaction, 77
  traditional functions, 77
Sensitive Urban Design (BSUD), 247
Separate patches, 226, 228, 229
Setback, 119
Shading coverage, 261
Shared street, 183
Shared transport, 198
  accessibility, 197
  affordability, 197
  benefits, 203
  bicycles, 203
  bikes, 199
  definition, 197
  digital platforms, 199
  dockless system, 201
  electric bikes, 198, 199
  enforcement, 201
  geo-fencing technologies, 201
  infrastructure, 200
  local governments, 200
  parking, 203
  parking stations, 201
  planners, 199, 200
  public and private transportation, 197
  public mobility, 198
  public transit, 198
  public transportation, 197, 200, 202, 203
  road networks, 200
  scooters, 198
  small vehicles, 200
  social and economic benefits, 198
  technology, 197
  technology-enabled mobility services, 198
  transport-policymakers, 199
  urban areas, 199
  urban mobility, 198
Shrub height, 259
Shrubs, 258
Sidewalks, 174, 175, 177, 210
Sightlines, 288, 292
Single-family homes, 92
Single-parent families, 98
Slateford Green, 185, 186
Small-scale variables, 117
Smart grid systems, 138
Smart growth, 45, 106–108
Smart home technology, 138
Social capital, 286
Social cohesion, 86, 93, 98
Social equity, 15
Social infrastructure, 6
Social interaction, 71, 73, 175, 213, 214, 279, 280, 285, 286
Social isolation, 207, 285
Social sustainability, 16, 286
Soil and rock formation, 237, 241
Soil erosion, 60
Solar power, 129
Solar-powered community, 13
Solar water heaters, 130
Spatial organization, 89
Spatial quality, 116
Spontaneous settlements, 38
Stormwater management, 252, 253, 279
Street design, 115, 116, 187, 190
Street furnishings
  built environment of cities, 301
  culture, 295, 297
  designers, 303
  implementation, 301, 303–306
  infrastructure, 300
  materials, 302
  microscale, 301
  placement, 302
  public art, 297, 298
  selection, 301
  texture, 303
Street-scale, 190
Suburban development, 45
Sustainability, 285, 286
  continuity, 24
  culture and governance, 16, 17
  economic, 18
  environmentally sustainable, 19
  governance, 19
  landscaping, 257

least negative impact, 21
life cycle approach, 24
placemaking, 264, 266
planting, 257
principles, 16, 21
self-sustaining system, 22
social, 16
symbiotic relationships, 23
urban design, 21
weather control, 258, 260, 262
xeriscaping, 262–264
Sustainable community, 34
Sustainable development, 15, 21, 51
Sustainable master plan, 33
Sustainable mobility, 148
Sustainable planning
economic and social activities, 3
economy and social life, 6
fundamental concept, 6
medieval cities, 3
mixed-use areas, 3
mixed-use public spaces, 5
modernism and unsustainable practices, 6
Sustainable systems, 21
Sustainable urban life, 89
Sustainable urban plans, 48
Symbiotic relationships, 23

T
Third places, 287
Topography
climates, 60
ecosystem, 57
environmental factors, 61
flora and fauna, 59
natural setting, 57
solid foundation, 57
Traditional high-density environments, 118
Traditional urban environments, 116
Traffic congestion, 6
Traffic congestion signals, 171
Transit-adjacent developments (TADs), 158
Transit apps, 138
Transit networks, 194
Transit nodes, 89
Transit-oriented development (TOD), 90
accessibility, 158
active mobility systems, 158, 160
advantage, 156
cost-effective transportation, 157
housing, 157, 160
land-use policies, 158
mixed zoning, 161
multimodal methods, 157
physical design, 155
planning, 160, 161
private vehicles, 155
public transit infrastructure, 155, 158, 160
strategic urban approach, 157
streetcar, 156
transit systems, 161
transportation system, 155
urban design, 159
Transportation, 187
planning, 151
policy, 157
Turf area, 263
Typology, 287, 288

U
Urban agriculture (UA), 277–282, 284
Urban coherence, 122
design, 115
environment, 115
neighborhoods, 116
small-scale variables, 116
Urban conservation, 120
Urban density, 226
Urban design, 83, 130
active mobility, 174, 177
active transportation, 172
benefits, 171
bicycle freeways, 177
bike rental stations, 178
cycling, 171
cyclists, 172
footpaths, 175
hierarchical system, 172
intersections, 176
maintenance, 175
parking, 174
pedestrian systems, 174, 177
physical interventions, 173
planners, 171
private vehicles, 171, 173
public transportation, 177
racks, 178
reallocating space, 172
road user charging, 173
shelters, 176
sidewalks, 177
strategies, 177
street furniture, 175
traffic congestion signals, 171
transportation, 173, 175, 179
urban landscapes, 174
vehicular traffic, 172
walking, 171

Urban design methods, 238–243
Urban design strategies, 214, 220
Urban developments, 13, 25
Urban fabrics, 71, 72
Urban food systems, 277, 279
Urban growth boundary (UGB), 107
Urban Heat Island (UHI), 245, 249, 258, 259
Urban IoT project, 138
Urban places' forms
  circulation designs, 43
  design itself, 40
  geometrical/orthogonal principles, 39
  human environments, 41
  human settlements, 37
  land-use planning, 43
  planning practices, 40
  Roman origins, 37
Urban planning techniques, 44, 150
Urban sprawl and single-family home, 4, 7, 24, 85, 95

V
Vanpooling, 198
Vertical layering spaces, 87
Vistas, 82, 83
Visual corridors, 82
Visual order, 116

W
Walkability, 18, 149, 158, 161
Walkable communities, 122, 268
Walking, 188, 190, 193, 213
Waste collection and recycling, 11
  biomass energy, 131
  energy conservation, 133
  industrial design, 133
  liquid sewage, 131
  water-usage, 132
  zero waste communities, 132
Waste disposal, 11
Water-use zones, 263, 264
Wayfinding, 210
Wealth, 207
Weather control, 258–260, 262
Wellbeing, 225, 226
Well-proportioned street width, 79
Wet functions, 113
Wetlands, 260
Wheelchair friendly, 270
Wi-Fi signals, 138
Windbreaks, 258
Wind flows, 64
Windrow of trees, 259
Winter cities, 220
  active transportation, 217, 218, 220
  challenges, 213, 214
  scope, 213, 214
Woonerf, 173
World Commission on Environment and Development (WCED), 15
World War II, 79, 80

X
Xeriscaping, 262–264

Z
Zoning policies, 7

GPSR Compliance

The European Union's (EU) General Product Safety Regulation (GPSR) is a set of rules that requires consumer products to be safe and our obligations to ensure this.

If you have any concerns about our products, you can contact us on ProductSafety@springernature.com

In case Publisher is established outside the EU, the EU authorized representative is:

Springer Nature Customer Service Center GmbH
Europaplatz 3
69115 Heidelberg, Germany

Batch number: 10756376

Printed by Printforce, the Netherlands